CW00613511

INTRODUCTORY TECHNOLOGY

INTRODUCTORY TECHNOLOGY

A Resource Book

ADRIAN OWENS

Drawings by Richard Burton
and Richard Mitchell

VSO/IT Publications 1990

VSO
317 Putney Bridge Road
London SW15 2PN, UK

Intermediate Technology Publications Ltd
103–105 Southampton Row
London WC1B 4HH, UK

© VSO 1990

ISBN 1 85339 064 X

Typeset by Inforum Typesetting, Portsmouth
Printed in Great Britain by
Short Run Press, Exeter

CONTENTS

ACKNOWLEDGEMENTS

The author would like to thank everyone who contributed ideas and advice during the writing of this book. Unfortunately it is not possible to mention everyone by name. But special thanks must go to the following: VSO London, Robin Story, John Fielding, Joy Voisey, Claire Wright, Martin Kyndt, Sara Ryan, Simon Lilly, Louise Haycocks, Ben Drew, Colin Dicks, Tim Lewis, Gary Morter, Dick Sharples, Carol Lee, Christine Serle, Richard Burton, Richard Mitchell, Simon Kellard, Robin Laney, Billy MacInally, Nancy Haworth, Florence N. Emeruwa, Gina Carver and Dave Harris.

FOREWORD

Introductory Technology is learning by doing, so that students can discover that technology is not something magical but something that is all around them, from their farming tools to the aeroplanes that may pass overhead. The aim of this book is to illustrate methods of teaching, ways of using materials and how to create an enjoyable and efficient teaching and learning environment; but most importantly, to show that, whatever difficulties the teacher may face, it is always possible to teach.

When teachers first arrive at their school there may be few facilities for teaching the subject; this should not stop the teacher from teaching. Start with what you have even if it is only a floor and a few tools. You must adapt your teaching to what is available but at the same time try to get better tools and facilities. It is hoped that a standard approach will be created, stimulating teachers to develop their teaching further by using the wide variety of skills, materials and ideas in current usage in their local community.

This teacher's resource book will be relevant more widely than in Nigeria, where it originated. The (6.3.3.4) system of education used in Nigeria is similar to much of West Africa, and similar courses are being introduced in Ghana and The Gambia, while the deficiencies in the curriculum which the approach described here seeks to address is of major importance to many countries.

'Introductory Technology', or 'Introtech' as it is known, is Nigeria's attempt to provide its young people with the technical 'life skills' which are increasingly necessary in all countries. This is not vocational training. It is attempting to provide skills which all young people need and which provide the necessary preparation if students are to make best use of further technical or vocational training.

VSO has been involved with Introtech from its inception in 1985. VSO Nigeria holds regular teacher training courses in the subject and provides VSO volunteers to work in demonstration schools in states in Nigeria. We believe the Nigerian government's initiative is an important one and this handbook will, we hope, provide another useful element in the development of Introtech.

NEIL MCINTOSH

Director of VSO 1985–90

VSO'S INVOLVEMENT WITH INTRODUCTORY TECHNOLOGY

In 1985, VSO was asked by the Federal Government of Nigeria to give special priority to assisting with the teaching of introductory technology, known as Introtech, in Junior Secondary Schools, and with the training of Introtech teachers.

VSO Nigeria holds regular introductory technology teacher training courses for both VSO teachers, their Nigerian colleagues, and ministry officials. The aim of these workshops is to train teachers in the methods of teaching that are needed for introductory technology.

The present aim of VSO Nigeria is to concentrate its resources over a fixed period of time in demonstration schools. These schools tend to have a high profile within a state so it will be possible for the school to become an Introtech resource centre which other teachers will be able to visit and see how the subject is being taught. In addition to the Educational Resource Centres and teacher training colleges, these centres will also be used for holding teacher training workshops.

VSO wholeheartedly supports the Federal Government's aims and objectives for Introtech, and has committed further resources to this part of their programme.

1. INTRODUCTORY TECHNOLOGY

For the development of any society there has to be a solid technical foundation on which the society can build. To live in such a society, the population must become technically literate. It is for this reason that introductory technology was introduced into the Nigerian education system, and similar subjects are now being introduced into many other countries.

The aims of introductory technology are:

- to provide basic technological literacy for everyday living;
- to build on traditional and local technology;
- to stimulate creativity by nurturing an open and questioning mind;
- to develop a respect for manual skills;
- to provide pre-vocational orientation for further training in technology.

SUBJECTS INCORPORATED WITHIN INTROTECH

The basic subjects incorporated within Introtech are:

- Metalwork
- Woodwork
- Mechanics/Physics
- Technical drawing
- Electrical
- Building
- Materials
- Local crafts
- Mathematics

NB: conservation should be emphasized throughout the teaching, especially that of natural resources.

INTRODUCTORY TECHNOLOGY AS AN INTEGRATED SUBJECT

Introductory technology should be taught in an integrated manner. This means that all the subjects incorporated within Introtech should be taught in such a way that students realize that technical subjects are all interconnected. For example, the making of small clay items can cover materials, local crafts, mathematics, building (making a basic kiln), drawing, physics (heat energy), woodwork (the making of brick moulds) and metalwork (the making of small tools to decorate the pottery with). The project can even be included in geography, history and English lessons. This is the type of approach the teacher should be thinking about and not only is it an effective method of teaching, it is also very enjoyable.

Because of the integrated nature of introductory technology it is very important that teachers work closely with members of other departments. Many of the problems faced, in terms of the quantity of work to be covered within the syllabus, may be overcome by coordinating schemes of work with teachers of related subjects. For example, forms of energy are covered within Integrated Science, so it may be possible to work with the teachers from this department when writing a scheme of work.

Teachers from 'non-technical' departments should also be consulted. One of the biggest problems faced by students is learning all the new words they will have to use. This can be partly overcome by the English teacher introducing the words into the English lessons. For example, when students are learning simple sentence construction, sentences can be based around a technical subject, e.g. 'Imo is cutting the wood with a saw'. The same can also be done in the maths lessons. By doing this, the student will soon become familiar with many of the new words and phrases.

MACHINES AND INTROTECH

After assessing the students' knowledge during the first lesson (see Chapter 8) the teacher will soon realize the basic nature of Introtech. In the past, the level of Introtech has been confused due to the installation of large machinery. This machinery is *not* needed for the teaching of Introtech but may be useful for the teacher in the preparation of materials, renting to local craftsmen and for senior secondary schools.

CREATING A STANDARD APPROACH

Despite the wide range of backgrounds and skill areas of teachers, the importance of a standard approach cannot be over-emphasized and there are four important points that, if followed, will assist in creating a standard approach and will help the teacher to teach the subject effectively:

1. Introductory technology must be taught through activities, so that students learn by doing.
2. Introductory technology is an integrated subject and must be taught as such for it to be effective.
3. Indigenous technologies should form the basis for introductory technology. The materials are easy to find as local crafts are built around available materials and an integrated project based on the technologies already present can act as a 'springboard'

for the rest of the syllabus. For example, the construction of a traditional bellows forge relies on the majority of disciplines within Introtech and demonstrates clearly the inter-relation of many technical subjects.

4. Introductory technology is pre-vocational. It is not aimed to produce craftsmen/women, but rather 'to provide technological literacy for everyday living', while at the same time giving students who want to enter technical education a clearer insight into the subject they wish to pursue.

2. SCHOOL ORGANIZATION

The aims of this chapter are:

(a) to explain the Nigerian 6.3.3.4 system of education – similar to much of West Africa's – and the role played by Junior Secondary Schools (JSS) within this system;
(b) to explain the internal organization of JSS and the duties the teacher will be expected to carry out.

6.3.3.4 SYSTEM OF EDUCATION

6 This represents six years of primary education, at the end of which successful students receive a Primary Leaving Certificate which is needed for entry into secondary education.

3 Three years of Junior Secondary School (JSS), which is both pre-vocational and academic. (Every student offers twelve subjects.) Introductory technology is taught in JSS years 1, 2, and 3, at the end of which the student will sit an external exam (see Chapter 8). The successful students receive the Junior School Certificate (JSC).

3 Holders of the JSC may proceed to the second stage of secondary education. There are three types of institution at this stage: Senior Secondary Schools, Technical Schools and Teachers' Colleges. At the end of three years the student will sit for the Senior School Certificate (SSC).

4 Students who have the SSC may attend either a university or polytechnic for further studies. Most basic courses run for four years.

Students who decide not to continue to Senior Secondary School may gain entrance into Vocational Technical Colleges. The aim of the VTC is to produce skilled craftsmen/women. At the end of three years the student will take trade test examinations.

SCHOOLS' INTERNAL ORGANIZATION

ADMINISTRATION

The administration duties of the school are performed by the Principal and the Vice-Principal (VP). They are responsible for the everyday running of the school.

SCHOOL FEES

In Nigeria, most students have to pay some of their fees at the beginning of each term. Once collected, the fees are handed over to the Ministry of Education. The Ministry then gives a percentage of this back in the form of an imprest which is then used for the running of the school.

PARENT–TEACHER ASSOCIATION (PTA)

Most schools have PTAs, the aims of which are:

- to raise money for the improvement of the school;
- to help maintain the educational standard of the school by holding regular meetings regarding the progress of their sons and daughters.

Many schools have a PTA levy that has to be paid along with the school fees. Through this levy many PTAs have been able to buy school buses, build workshops and install equipment (see Chapter 11).

BOARD OF GOVERNORS

The Board of Governors is made up of distinguished people from the community – often ex-students. The Principal is responsible to the board, and it is through the board that major decisions concerning the running of the school are taken. The board holds regular meetings but can also be called upon if the need arises.

TEACHERS' ADMINISTRATIVE RESPONSIBILITIES

Apart from the usual responsibilities expected from the teacher such as marking and recording, there are two other very important duties that must be carried out.

FILLING IN OF SCHOOL DIARY

Every school will have a diary that has to be filled in at the following times:

- at the beginning of each term to show the scheme of work being used for that term's teaching;
- at the end of each week to show what lessons have been taught.

It is most important that the diary is filled in as it will prove invaluable to incoming teachers. It also gives the serving teacher a clear picture of how the students are progressing – most important with a new subject such as Introtech. The diary is usually kept by the year supervisor and it is up to the teacher to *ask* and find out where the diary is kept.

INSPECTION OF LESSON PLANS

Each week the teachers will be expected to pass their lesson plans to the head of department for inspection. The head of department will in turn pass theirs to the Principal or VP.

It is particularly important that introductory technology teacher's lesson plans are checked to ensure that the integrated nature of the subject is being followed.

TIMETABLE

It is imperative that enough periods are allocated to the teaching of introductory technology. The recommended minimum number of periods per week is five 40-minute lessons.

As introductory technology is a new subject many school administrators are not aware of its timetable needs and the teacher will have to tell the timetable committee what periods are required. Before teachers can do this they must form a clear picture in their minds of how they are going to approach the subject. There are a number of determining factors that will affect their approach and therefore the timetable needs. The following should be taken into consideration before deciding on a timetable format:

- number of students per class;
- number of classes per year and number of years to be taught;
- number of teachers;
- scheme of work and syllabus to be followed;
- facilities available;
- method of teaching, i.e. how many double periods will be needed.

Once the above information has been collated, a decision can be made on the most effective timetable format. However, it is necessary to bear in mind one other outside influence – timetable pressure! Every other teacher in the school is also looking for a perfect timetable, so in the end a compromise will have to be arrived at. Although the timetable may be fixed when you arrive at your school, it is always worth trying to get it changed, especially if you feel it is affecting the quality of the subject.

DOUBLE PERIODS

Double periods are essential for teaching activity lessons. Two doubles and one single period a week is ideal. The single period can be used as a summary of the week's work and for giving any back-up information that is needed.

The most important point to remember about the timetable is that it must be *manageable*. To teach six classes with a well-planned timetable allowing the teacher ample preparation time is better than trying to teach twelve classes and having little time to prepare – overloading oneself will only be counter-productive.

AFTERNOON/EVENING CLASSES

See Chapter 6.

3. THE INTRODUCTORY TECHNOLOGY WORKROOM

It doesn't matter whether there is a workshop already set up or a classroom, an environment conducive to the teaching of the introductory technology can be set up anywhere – even outside under a tree! The aim is to create a learning environment that is exciting and will stimulate creativity, encouraging students to think and discover things for themselves. This chapter explains how such an environment can be created. For convenience the word workroom has been used to describe this environment.

WHAT IS A WORKROOM?

The workroom is a joint classroom and workshop where projects and activities are carried out. You don't need both benches and desks; if you have benches they can be used as desks, if you only have desks then improvise – floors provide a good working area! The rest of the workroom should be full of interesting stimuli such as posters, mobiles, resource areas, etc. (Fig. 3.1). Listed below are many teaching aids that help to create both an interesting environment for the students and an efficient system of working for the teacher.

DRAWING EQUIPMENT CUPBOARD AND BOOKCASE

A small cupboard fixed to the wall containing all the equipment needed by the students for drawing, such as compasses, pencils, rulers, and coloured pens, is very useful. The cupboard can also contain textbooks, worksheets and question-bank cards (see Chapter 7). The front of the cupboard can either be small wire mesh so it's easy to check without opening, or a solid door which can double as poster-board/display area. The cupboard should be kept locked and a simple/quick checking procedure devised. A cupboard monitor could be chosen from each class to be in charge. An old desk top can easily be converted into a good cupboard (see Materials, Chapter 12).

SAMPLE TOOL BOARDS

It is very important that the students become familiar with the names and functions of tools used in introductory technology. A

A bad working environment

A good working environment

Fig. 3.1 Working environments

very good method of doing this is by making sample tool boards which may contain one of each type of tool (if available), or may just be drawings of the tools. These boards can then be hung around the room. If the labels are made in such a way that they can be removed, the boards can be used for quick 'fill in' revision exercises.

CHALKBOARDS

If the workroom does not already have a chalkboard the following two designs may be found useful:

1. Hinged chalkboard (Fig. 3.2): this gives a large work area and is very useful when several diagrams or project drawings need to be shown. It saves a lot of time with preparation and allows full lesson time to be utilized as no time is wasted producing drawings. Also, if the teacher has prepared work on the board for the students but doesn't want them to see it till later, it can be covered by the hinged board.

Fig. 3.2 Hinged chalkboard

2. Easel chalkboard (Fig. 3.3): very cheap and very versatile. Work can be prepared in advance for later use thereby saving lesson time.

Fig. 3.3 Easel blackboard

PICTURE AREAS

A small area can be set aside for sticking up pictures of technology. Students can be asked to find pictures showing technology, for example cars, oil rigs, local crafts, which can then be placed on a board or mat either at the beginning or end of each lesson after a short explanation from the students as to what the picture shows. It will probably be found that there is a large turnover of pictures, thus creating a new and interesting display each week. All old pictures should be kept as they can be used again, either for the board or workcards. Sources of pictures include papers, magazines, calendars and local paper bags made from newspaper.

NOTICE BOARD

A small area of wall can be used as a notice board.

WALL PAINTINGS

These are ideal for creating an enjoyable and interesting environment. They can range from small poster-size paintings to a large mural. These can be done by the teacher, the art department or a local artist. If you don't want to commit yourself to a permanent drawing, then do them in chalk using a charcoal/kerosene mix for a black background (see Chapter 12).

CEILINGS

Do not forget the ceiling. Mobiles of shapes, tools, etc. can be hung from the rafters.

MAGNETIC BOARDS

Many of us will already know about the use of the magnetic board – but most of us will not have one! There is no excuse! In most towns, schools, etc. there are broken fridges that can have many uses.

MAGNETIC BOARD

The front of the door can be used as a magnetic board. It is already painted gloss white and so can be drawn on with water-soluble pens. The magnetic strip inside the rubber surround inside the door can be stuck on to visual aids in order to fix them on to the board.

POLYSTYRENE BLOCK

The insulation material inside the door can be removed and used for models. This takes a lot of time and patience, cutting it into strips and levering if off.

BLACKBOARD

Plywood can be inserted into the back of the door and painted with blackboard paint. This gives you a portable magnetic board/chalkcard.

DISPLAY AREAS

Display areas (Fig. 3.4) are an absolute must for the workroom. It is important that the display is changed on a regular basis. If no tables are available, desks can be pushed together against a wall and covered with material or mats. Two areas should be set aside for displays.

STUDENTS' WORK

Examples of students' work should be continuously displayed. The teacher should change exhibits regularly to ensure each student's work is displayed.

GENERAL INTEREST

Displays may be linked with an on-going project or organized purely to create interest and provoke discussion.

Fig. 3.4 Display areas

WAYS OF DISPLAYING

When presenting a display it should be:

- visually attractive;
- interesting and easy to understand;
- relevant;
- carefully positioned so students take notice of it.

One interesting way of starting a general display is for the teacher to present one item which a display will be built around. For example, the teacher can start the display with a catapult and then ask the students to bring any items that use two or more materials. From the items brought, activities can be carried out to discover more about materials and their properties.

The following are a few more display ideas:

WOOD

A display may be started by using a large log and asking students to bring anything made of wood. From the items brought, the teacher and students can build displays around specific areas such as joints, parts of a tree, timber faults, papermaking, charcoal, etc.

MATERIAL SAMPLES

A display showing different materials and their uses. It is important to use items that the students will know, i.e. parts of machines, cars and bicycles for steel, aluminium and plastic.

SIMPLE MACHINES

Use a bicycle as the centre of the display identifying all the simple machines within a bicycle, i.e., levers (brakes, pedals, etc.), then display other items that use the same mechanical principles, e.g. pedals and farming hoes both use the principles of levers.

LOCAL CRAFT DISPLAY

Show examples from all local crafts.

POTTERY

Local clays and additives used, methods of making and firing, uses of pots, i.e., toys, ceremonial, cooking and water storage.

METALWORKING

Examples of all the metalworking crafts showing materials, tools, methods and uses of the finished product.

CALABASHES

Where they come from, uses, tools used for decoration and techniques used.

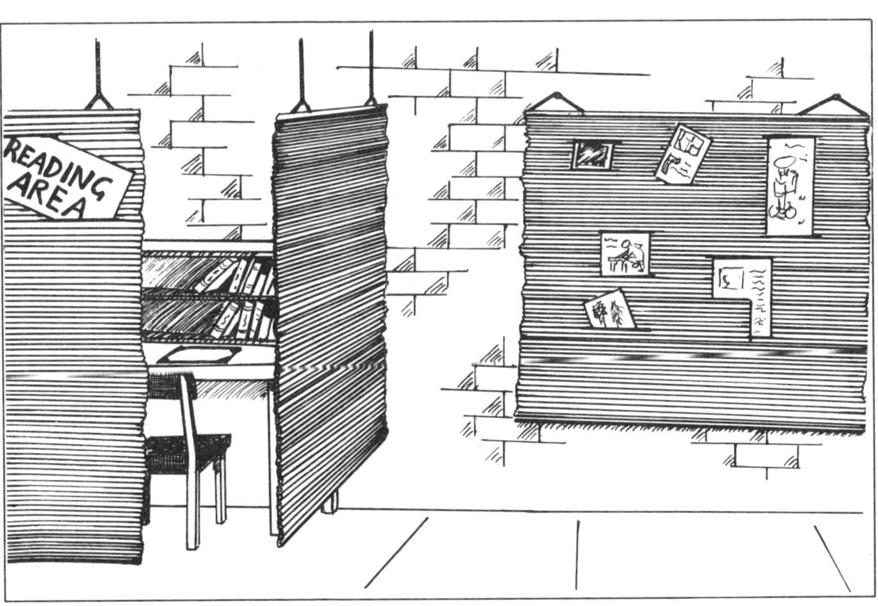

Fig. 3.5 Mats and wall-hangings

MATS/WALL-HANGINGS

There is no need to paint all the walls; just hang large cornstalk mats on the wall and fasten posters to them. If you want to divide an area of the workroom off for a quiet area you can hang mats from the ceiling which double as display areas (Fig. 3.5).

TEACHER'S DESK

It may seem obvious but is important.

LIBRARY/RESOURCE AREA

A small area of a desk, wall cupboard or floor can be set aside as a resource area (Fig. 3.6) which may include worksheets, question banks, technical library, pictures and models (see Chapter 7). By setting up a resource area there should always be 'fill in's' for the teacher and work for the students.

Fig. 3.6 Resource area

4. SETTING UP THE WORKROOM

This chapter deals with the construction of workroom equipment, inventories, methods of storing tools efficiently and the maintenance of tools. When designing and making equipment it is important that you think carefully about the final style of the workroom and use the available space wisely (see Chapter 3).

INVENTORIES

The first thing you must do on arrival at the school is to write a detailed inventory. If there is one already, you *must* check it. Not only should an inventory contain all the tools and equipment but also all the spares. Once written, copies should be given to the Principal, storeman and other departments.

If you are just in the process of setting up your workroom and you know teaching will not begin for some time then it is a good idea to remove the tools you will need and pack the rest up until your students are ready to use them.

WORKBENCHES/TABLES

Workbenches seem to be a stumbling block for a lot of teachers. Obviously the teacher will require some money with which to buy materials, but it is surprising what materials can be found for nothing. The benches shown here vary from very simple and very cheap to the traditional type of workbench. The style chosen will depend on the resources available and the work to be carried out.

PACKING CRATE BENCH

If your school receives a consignment of tools or machinery it will probably arrive in large packing crates. These crates are an excellent source of wood for building benches – don't throw away the metal straps that hold the crates together; they can be used for making small knives and other tools.

WALL BENCHES

A good space-saving style of bench is the wall bench (Fig. 4.1). At its simplest this may be a board of wood that runs along the wall (inside or outside). It may be fitted with hinges to save space. (Although at first it seems that there is not much working area for the students, it must be remembered that the majority of the students' projects will be fairly small.) It is also very useful as an extra working area.

Fig. 4.1 Wall benches

STANDARD BENCH

This is the traditional bench, although the styles vary according to the maker and materials available (Fig. 4.2).

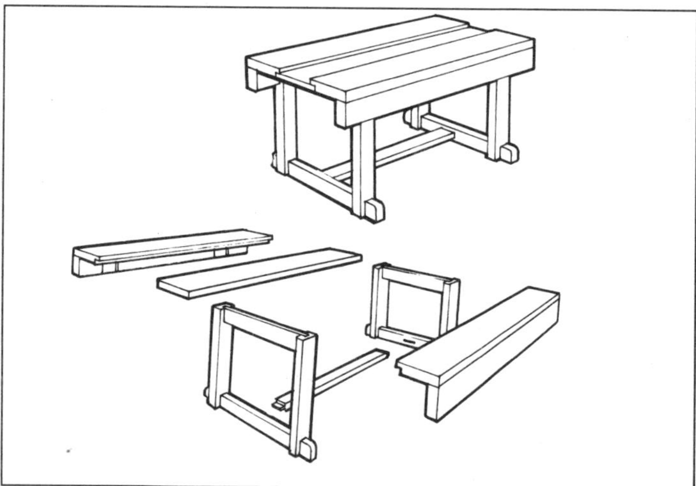

Fig. 4.2 Standard-type bench and its component parts

MINISTRY BENCHES

Some education ministries supply very good workbenches. The teacher should find out if these benches are available to them before they start spending time and money constructing benches.

A vice can be *anything* that holds the work (watch the local craftsmen at work and study their vices!). When you need a work-holding device it is important to think simple; forget about the designs you are used to and look at what is available.

F CRAMPS AND G CRAMPS

These can be used in many different ways to hold work. By using them in conjunction with pieces of wood and metal many excellent work-holding devices can be made (Fig. 4.3). Note that the back of the cramp can be bolted to the rear of the bench.

Fig. 4.3 F cramps and G cramps

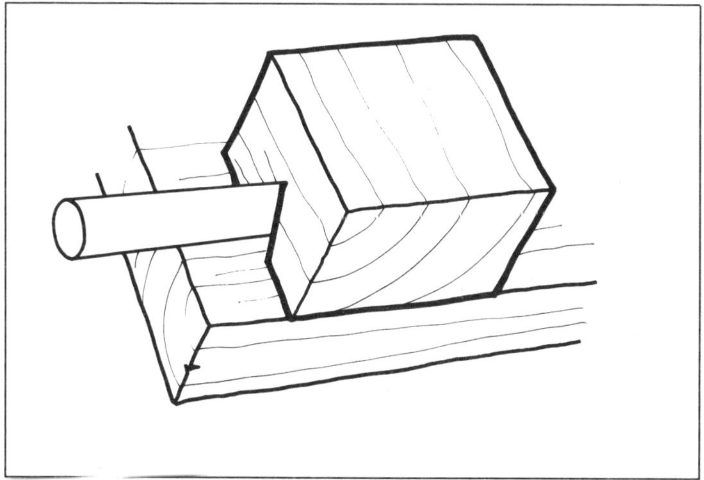

Fig. 4.4 V blocks

V BLOCKS

Different sizes of wooden V blocks can be made and used in conjunction with F cramps (Fig. 4.4). It will probably be found that only three different sizes will be needed, as the sizes of round bar available are limited.

ANGLE-IRON VICE

A very simple, cheap, but effective vice can be made from two strips of angle-iron (Fig. 4.5). The wooden V blocks can be used in conjunction with this in order to hold round bar.

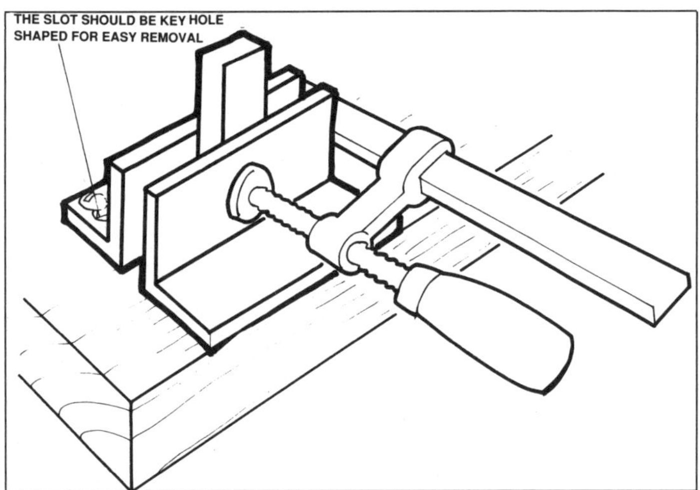

Fig. 4.5 Angle-iron vice

It can be seen from the above ideas that your whole work-holding system can be based around F cramps and G cramps. You can develop and refine your own ideas for holding devices using F cramps, and you may find your designs work better than a 'real vice'.

LEG VICE

A very effective vice which can be built into your workbench design is the leg vice. However, it may be a bit cumbersome when working with a lot of students (Fig. 4.6).

WEDGE VICE

In many people's opinion the example shown (Fig. 4.7) is the simplest, cheapest and most effective of the wooden wedge vices and can easily be fitted to any type of bench or table.

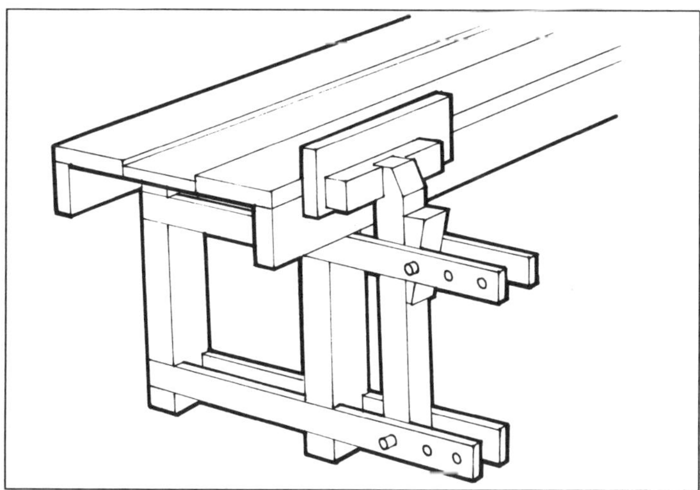

Fig. 4.6 Leg vice

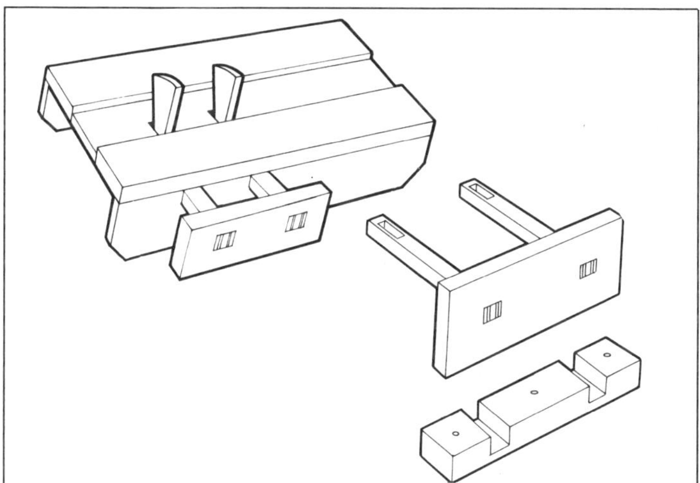

Fig. 4.7 Wedge vice

STOPS AND WEDGES

By visiting the local craftsmen you will soon see many very effective holding techniques, using a combination of stops and wedges.

FORGES

Forges are an invaluable piece of equipment for teaching introductory technology for the following reasons:

1. They are cheap and simple to make.
2. They are traditional in most areas so students are familiar with them.

3. A great variety of work can be carried out on the forge – not only metal-working but many other processes needing heat, including science experiments.
4. They can be portable and used outside to save room.

The style of forges changes from area to area. In some areas the most common type is the skin or tyre bellows whereas in other areas forges are mainly based around fans. Remember to think simple when making a forge. Basically you need heat and air using the simplest method available.

PORTABLE BELLOWS FORGE

This is an extremely cheap but effective forge. The size of the furnace depends on the work to be carried out but can be as small as a large milk tin. The inside of the furnace can be lined with clay to prolong its life. When using clay as a lining, heat the clay very gently to avoid cracking. Skills involved in making a forge include

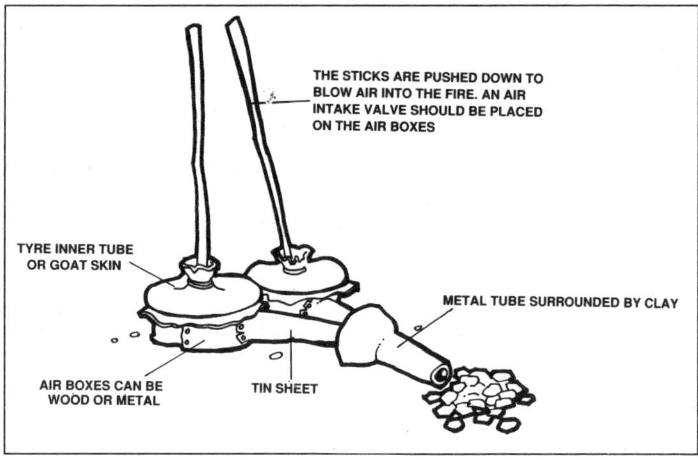

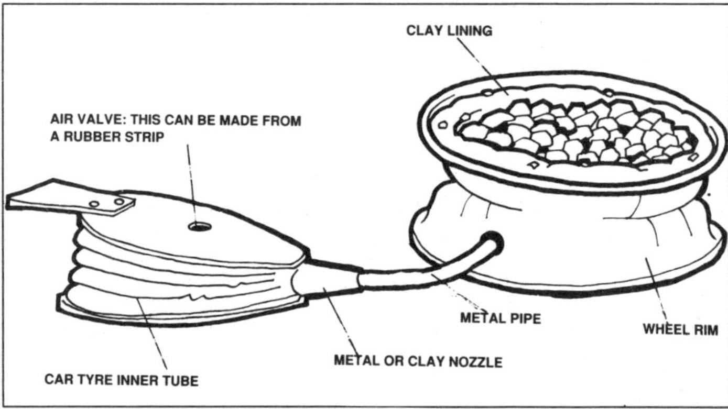

Fig. 4.8 Portable bellows

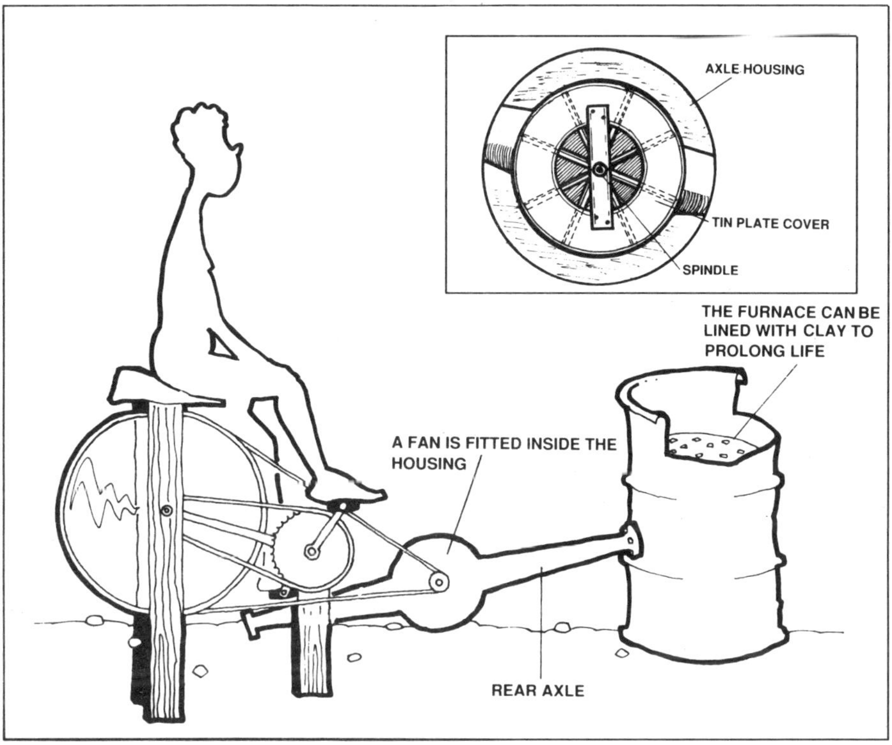

AXLE HOUSING
TIN PLATE COVER
SPINDLE
THE FURNACE CAN BE LINED WITH CLAY TO PROLONG LIFE
A FAN IS FITTED INSIDE THE HOUSING
REAR AXLE

Fig. 4.9 Fan-type forge

UPPER TIN HAS NO TOP OR BOTTOM
BAMBOO OR ANY HOLLOW PIPE
THE LOWER TIN HAS A TOP WITH MANY HOLES IN; THIS ALLOWS FOR THE AIR TO SPREAD EVENLY AND ALSO ACTS AS AN ASH PIT

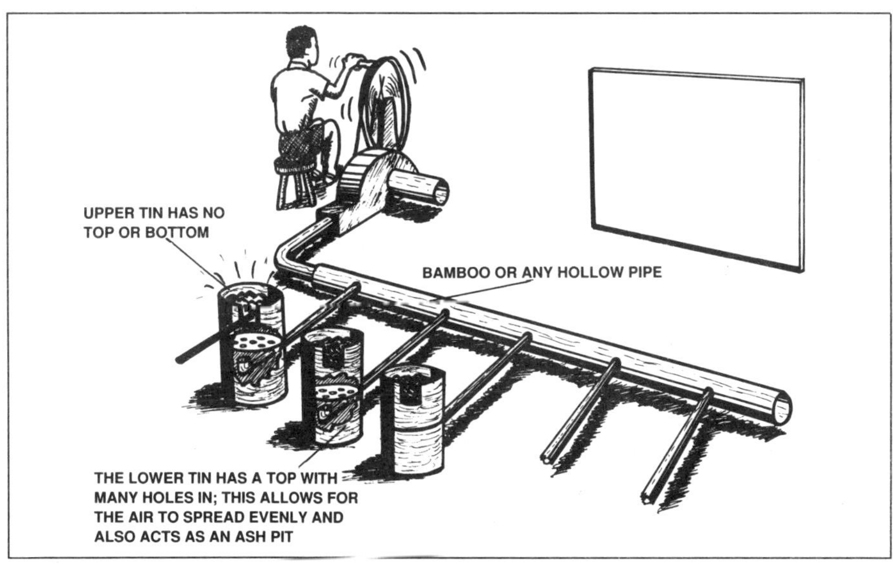

Fig. 4.10 Multi-forge

woodwork, metalwork, working with clay, cloth, and understanding heat energy (most of introductory technology in one project!) (Fig. 4.8).

FAN-TYPE BLOWER

The fan-type blower is slightly more complex than the bellows, but is still relatively easy to make. If the teacher wants to do some casting, then it is better to make a fan-type forge as it is easier and quicker for melting aluminium. The size depends on work to be carried out and can vary from milk tins to oil drums (Fig. 4.9).

If students are only doing small work many small forges can be run from one blower (Fig. 4.10).

None of these forges are perfect, and each of those illustrated has its advantages and disadvantages – but they all work and they are all makeable!

FUEL FOR FORGES

Charcoal is the most commonly used fuel in the forge and is readily available in most areas. It produces a fairly clean fire but burns quickly, therefore requiring quite large amounts if a lot of work is being carried out.

In some areas charcoal is either hard to find or very expensive. A cheap and very effective alternative can be found in palm kernel shells, from the palm oil tree. The kernels are pressed to extract the palm oil and the shells are usually discarded. Before using the shells they should be dried in the sun, otherwise they produce a lot of heavy smoke. Although once dried they can be used 'straight' it is better to burn them first in a bin to produce charcoal as this gives a nice clean fire.

NB: coconut shells can be broken and used in the same way as palm shells.

ANVILS

When looking for an anvil, again think simple – what type of work will you be producing? Once you have decided this, you should collect or produce large pieces of metal of the required shape. The following metals have proved very efficient:

1. Railway line – very nice if you can get it but not always readily available. Try the local Ministry of Works.
2. RSJ girder – usually more readily available than railway line. Try construction companies.
3. Caterpillar track – very good and already formed into a very useful anvil. Try asking at the Ministry of Works and road construction companies.
4. Engine mountings – large engine mountings from trucks are very good, especially for right-angled bends, and they also have

a variety of punch holes already present. Some even have a round bar welded on, which makes a very good bick.

Once you have found your anvil it is important that it is fixed *very* securely. It should not move when the work is struck by the hammer.

KILNS

Clay is a versatile and widely available material. Students can work on a variety of projects which encourage imagination and creativity at any level. If clay is locally available, a kiln is an excellent integrated project and a variety of different kilns can be made.

MOUND KILN

A mound kiln (Fig. 4.11) is a good design to start with. It is quick and easy to build and provides a way of firing students' work at an early stage. These are instructions on how to make the kiln:

1. Scoop a hole in the ground big enough to hold all the dry clay items.
2. Line with broken pots. (Visit a local potter for your first broken pots.)
3. Cover with dung, interspersing pots.
4. Cover with broken pots.
5. Layer sticks, small logs, dried grass and leaves, etc. Light from the top and allow to burn down.

Dung holds the heat and burns slowly, leaving a carbon deposit on the clay, giving the pots a black surface. Less dung and more wood between pots gives a clean, faster burn, revealing the natural terracotta colour. Beware of too fast a burn as the pots may explode!

Fig. 4.11 Mound kiln

If the teacher has access to sawdust a variation on the above kiln can be made as follows (Fig. 4.12):

1. Dig a small pit in the ground.
2. Lay three pieces of scrap metal tubing into the pit, pointing in different directions.
3. Kindle a small fire with paper and sticks in the pit base.
4. When the fire is well alight, add some sawdust; this will deaden the flames – smoke should still issue from the tubes, indicating that sawdust is still smouldering.
5. Add alternate layers of pots and sawdust.
6. Cover with original soil or turf, and allow to burn out overnight or for approximately twelve hours.

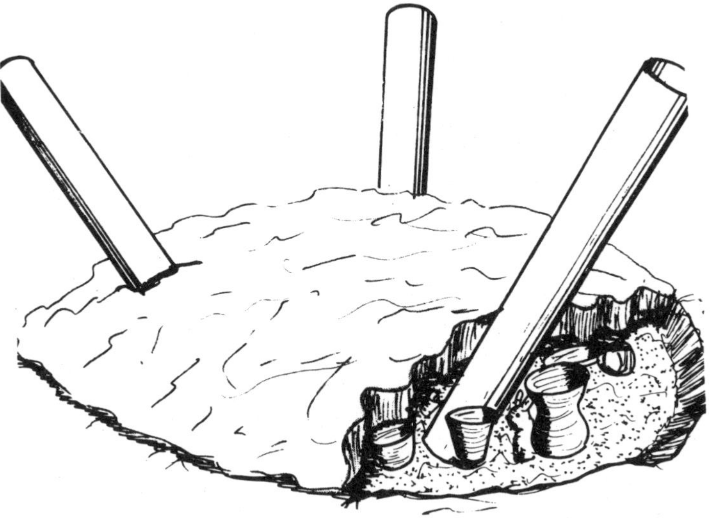

Fig. 4.12 Variation on the mound kiln

WALLED KILN

The simple mound kiln is most easily modified by adding a circular wall around it to hold the heat in on the sides (Fig. 4.13). This reduces the fuel needed and increases the temperature attainable. If you are making your own bricks for the wall (the cheapest way), students will lose interest if they have to make too many. Remember, the kiln doesn't have to be too big. Students will be far more motivated if they can see something happening; start small and, if it is successful, build a larger one.

Fig. 4.13 Walled kiln

STORAGE SYSTEMS

NUMBER OF TOOLS REQUIRED

The problem many teachers face when setting up a storage system for the first time is how many tools to include. What often happens is that too many tools are included and many are never used. Students are very happy to share tools such as hammers and centre punches that they will not be using all the time. Think carefully before deciding on the number of tools needed.

All spare tools can be oiled up and stored away until they are needed, i.e., when other tools are damaged or lost.

SYSTEMS OF STORAGE

Once you are ready to start teaching, you will need to decide what type of storage system you will use. These are suggested criteria by which to work when making the decision:

- Is it secure?
- Is it simple and quick to use?
- Is it simple and quick to check?
- Does it allow students freedom in choice of tools?

- ○ Is the system easily transferable to other teachers? Often storage systems are set up by one person without consulting other departments or considering the teachers that may follow. Do not set up a storage system that is only known to yourself. Make things plain and simple with no quirks.

SHADOW BOARDS (FIXED INSIDE STOREROOM)

Shadow boards are probably the most common type of storage system and, under the right circumstances, very effective; the outline of the tool is painted on the board or wall. The cheapest method of setting up your storeroom in this way is by putting nails straight into the wall and hanging the tools from them. (You will probably have to use a punch to start the hole.)

The usual way of using this system is by allowing students free access into the store. During the lesson no student is allowed out of the class without signing their name. At the end of the lesson the storeroom is checked. If the tools have been stored and labelled well, the store will be quick and easy to check. However, there are drawbacks to allowing students free access, which are as follows:

- ○ Students must have pre-knowledge of all the tools, otherwise a lot of time can be wasted – even if the tools are labelled.
- ○ The storeroom must always be very tidy with nothing lying around the floor or shelves – this is inviting someone to steal.

Always remember: a place for everything and everything in its place.

Alternatives to allowing all students access are:

(a) To appoint tool monitors, which is fairly effective but can be time-consuming for the monitor.

(b) To put a basic set of tools out before each lesson. This is a good method if individual tool-boxes or trays have been made which can be placed on each bench.

SHADOW BOARDS (MOVABLE)

Movable shadow boards (Fig. 4.14) save space and are quick and easy to use.

The style of board can vary according to materials available, the size of storage space, etc. but should follow these guidelines: the board should contain the basic tools needed; each board should hold enough tools for one bench and the board must be able to fit on or near the bench. A good method is to hang the boards on the end of each bench at the beginning of each lesson and remove them after checking. Each bench can have a tool monitor who is responsible for all the tools and the removal of the board at the end of the lesson. Because only basic tools are placed on the board, an extra tool cupboard is needed.

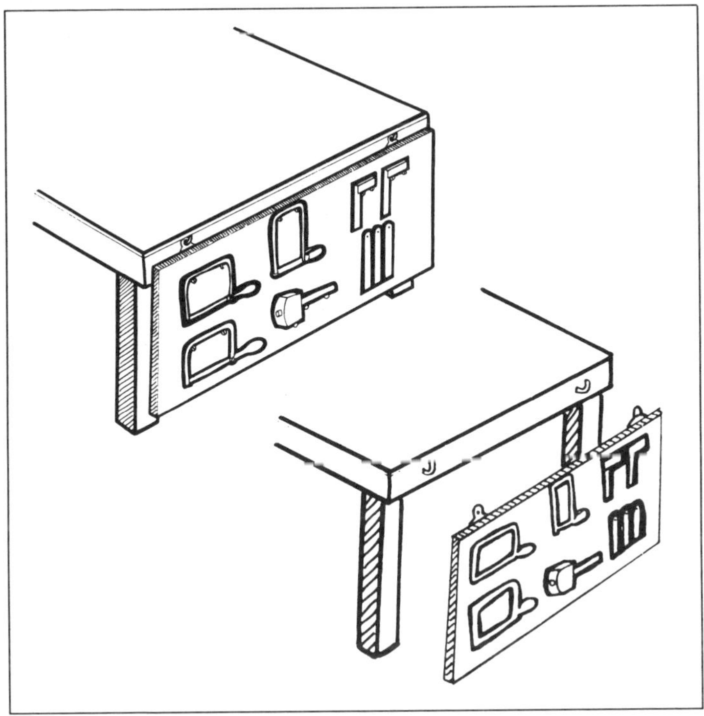

Fig. 4.14 Movable shadow boards

MOVABLE TOOL-BOXES

Tool-boxes (Fig. 4.15) should contain basic tools only and enough for each bench. The design of the box is very important. What may seem to be a clever space-saving design, may in practice present problems with removing, replacing and checking tools.
Remember there must be a convenient and safe place to put the box while the students are working. If a good design is used, this is an extremely efficient method of storing and using tools.

BENCH CUPBOARDS

This is a tool cupboard actually on the end of each bench (Fig. 4.16), containing a basic set of tools. When making a cupboard like this, be sure the tools are easy for the student to remove, with no danger of them hurting themselves on sharp tools.

EXTRA TOOL CUPBOARDS

If the storage system is based around movable shadow boards, tool-boxes or bench cupboards, a tool cupboard will be needed to

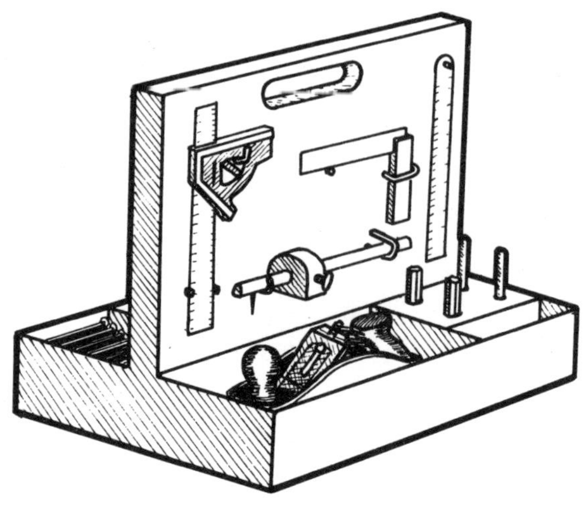

Fig. 4.15 Movable tool boxes

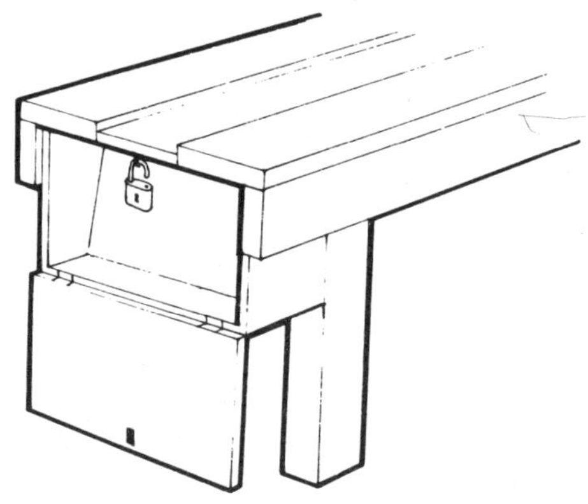

Fig. 4.16 Bench cupboard

store the extra tools that are required from time to time, i.e., drills, hand vices, etc. The cupboard can be an area of the storeroom or purpose-built in the workroom. Remember: it must be very organized and easy to use and check.

TROLLEY

An excellent way of storing tools is in a trolley (Fig. 4.17). Toolboxes, movable shadow boards, etc. can be stored on top of the trolley while extra tools can be placed inside.

The trolley needn't be too expensive to make. By looking in the school store or around the compound, an old desk (office-style) may be found that can be converted. Wheels can be made from hardwood.

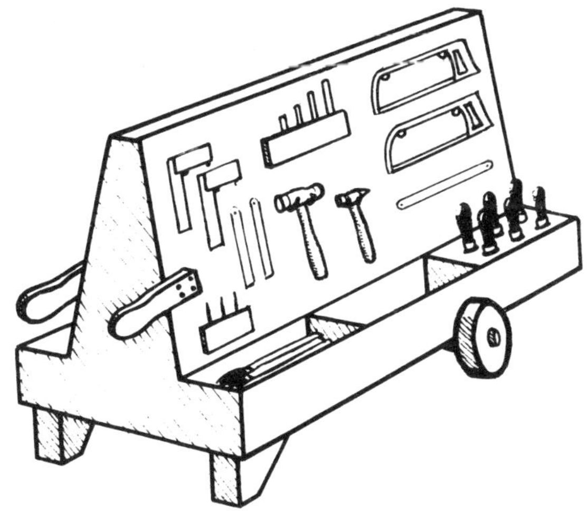

Fig. 4.17 Trolley

COLOUR-CODING

When using basic sets of tools for each bench, it is a good idea to colour-code them; then students can be organized into colour groups, i.e. the blues, greens, etc. This saves a lot of time starting and finishing lessons and also helps to solve the problem of students taking tools from other people's benches.

TOOL MONITORS

Appointing tool monitors gives responsibility to students and saves a lot of time at the beginning and end of lessons. When using tool monitors, remember the following:

- Be sure all students are aware who the monitors are (give monitors a special badge).
- Change monitors regularly so each student gets a chance.
- Be sure that the student doesn't lose working time by being a monitor.

SIGNING OUT BOOKS

From time to time school maintenance staff may ask to borrow tools for minor repairs within the school compound. (Never allow tools to leave the school compound.) All tools borrowed must be signed out and a return date agreed upon.

SECURITY

It is essential that the workroom and storeroom are very secure. If, when you arrive at the school, the storeroom is not secure, all tools should be moved to a secured place, e.g. the Principals' office. On arrival, carry out the following security checks:

1. Are the padlocks of good quality? Do not use cheap ones.
2. Is there a gap between the top of the workshop wall and the roof? Many workshops have been broken into by people climbing through this gap.
3. Are there security bars on the window?
4. If the storeroom has a window, are there shutters?
5. Is there a nightwatchman?

If security is not up to standard a meeting must be arranged with the Principal and, if it is a community school, the PTA and local community. Faults in security should be pointed out and it must be made clear that work cannot start until there is adequate security. It should be suggested the school nightwatchman patrols the workshop.

MAINTENANCE OF TOOLS

It is essential that all tools are kept in good condition. This may seem a very obvious statement but, with a hundred and one other things to think about, it is all too easily neglected.

By planning a simple maintenance schedule, tools can be kept in order with the minimum of fuss. One hour a week should be enough time to carry out maintenance.

It is important to involve the students as it is often during the maintenance and close study of tools that students really understand how they work, and preventative maintenance is an important part of Introtech.

SHARPENING TOOLS

To keep saws, planes, chisels and drill bits in good condition, they must be kept sharp. Often a teacher new to introductory technology may shy away from sharpening tools (especially saws), thinking that it is difficult – it is not and the teacher should learn how to keep tools sharp. Methods of sharpening saws, chisels, plane blades and drill bits are shown below.

PLANE BLADES AND CHISELS

Basically two processes are involved: grinding and sharpening.

Grinding If you do not have a grinding wheel, someone will. It may mean a trip into town once a month to have tools ground. Plane blades and chisels are ground at an angle of 25° (Fig. 4.18). If you are using an off-hand grinder, quench the tool very frequently in water to avoid it burning. NB, chisels and plane blades will only need grinding when the 25° angle becomes very short due to constant sharpening.

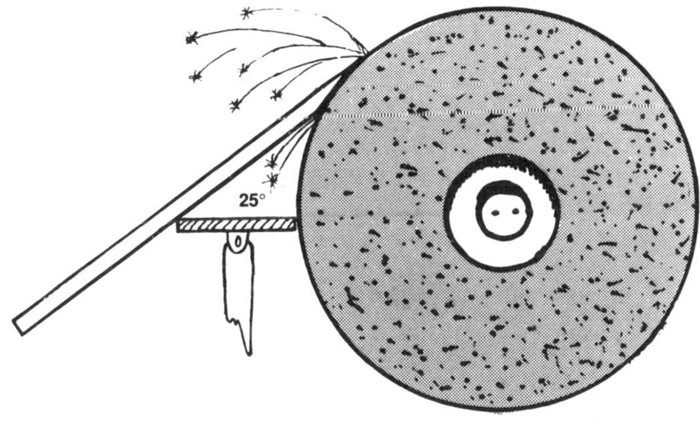

Fig. 4.18 Grinding

Sharpening This consists of putting another bevel at 30°, using an oilstone (Fig. 4.19). Put a small amount of thin oil on to the stone. To find your angle when sharpening, lift the blade until a line of oil is seen squeezing out between blade and stone. When you see this, raise the chisel again, very slightly (i.e. to 30°). Rub the blade up and down the stone in a figure of 'X' pattern. Continue the rubbing action until a burr (rough edge) is formed on the flat side of the

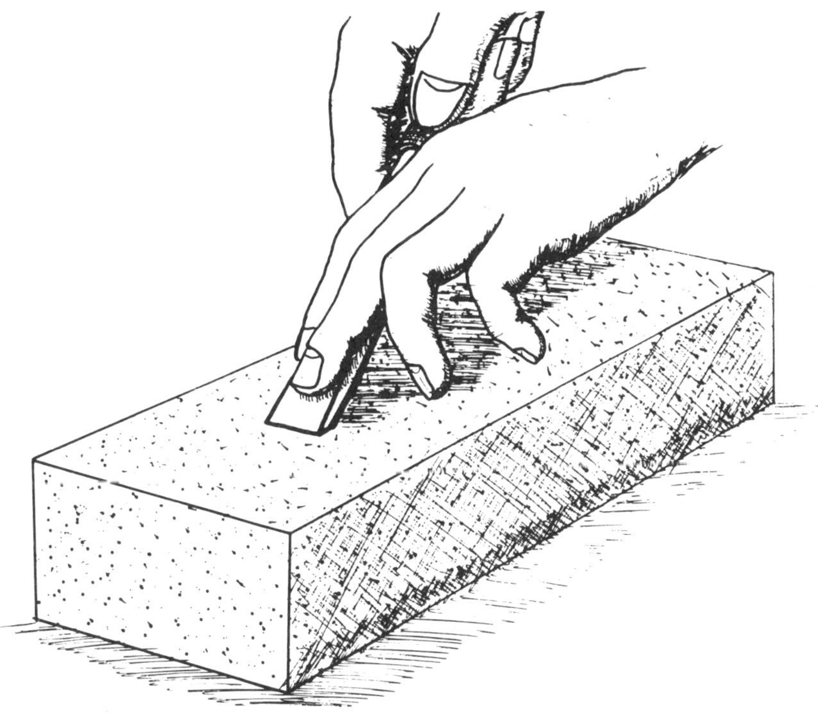

Fig. 4.19 Sharpening a chisel

Fig. 4.20 Sharpening a saw

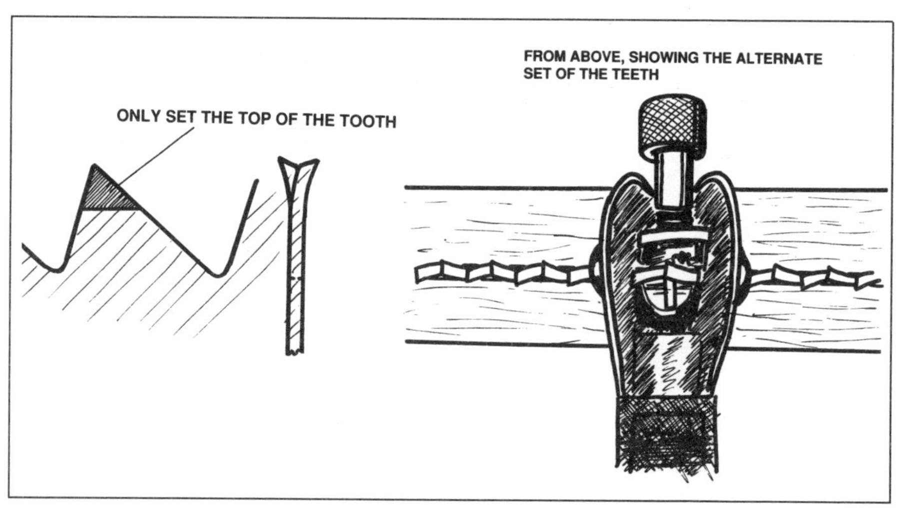

Fig. 4.21 Setting the saw

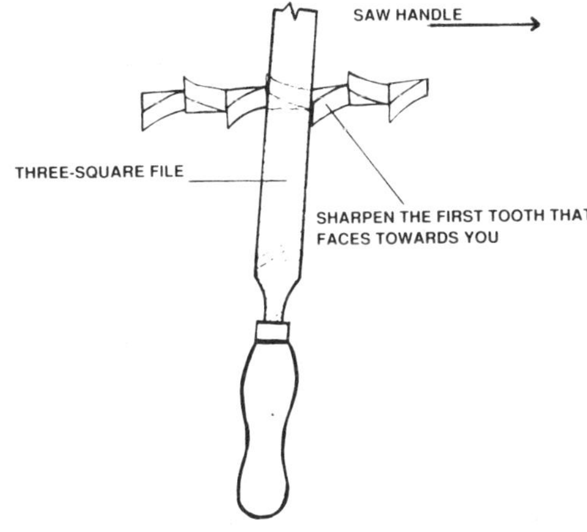

Fig. 4.22 Final sharpening

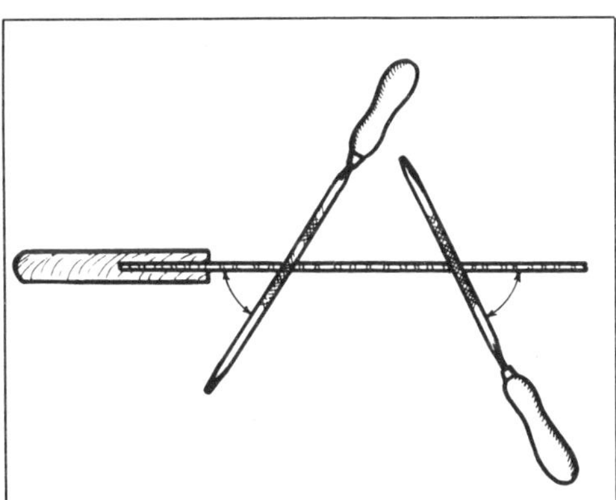

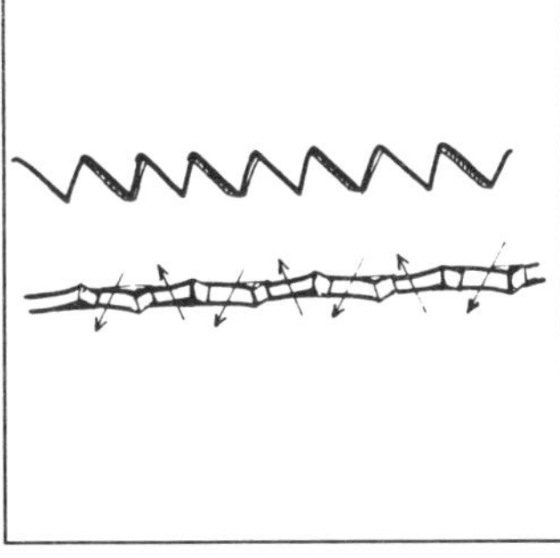

Fig. 4.23 Cross-cut saws

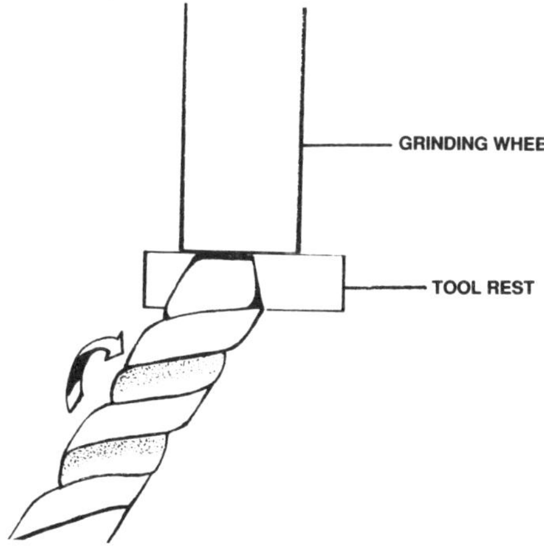

Fig. 4.24 Grinding drill bits

blade. Remove the burr by placing the flat side of the blade on the stone (it must be kept dead flat). If you don't have an oil-stone find out what local craftsmen use.

SHARPENING A SAW

When sharpening a saw it must be supported firmly. Hold the saw between two pieces of wood either in a vice or with G cramps (Fig. 4.20). You will need a small triangular file (twice the depth of the saw's teeth). If the saw is in poor condition, the following will have to be done before it can be finally sharpened:

1. File the top of all the teeth until they are the same level.
2. If the teeth are misshapened, file them back with a small triangular file.

Setting the saw (Fig. 4.21) Set the saw before final sharpening. The easiest method is to use a saw set (if you don't have one ask around). If a saw set is not available, place the saw flat down on a piece of soft wood and set the teeth by using a small punch (only the top of the tooth is set). The amount of set will have to be judged by eye from other saws. To check that all teeth are set the same, hold the saw at eye-level against the light and look for any unevenness.

Final sharpening Position the saw so the handle is on your right and put the file on the front edge of the first tooth set towards you (Fig. 4.22). Keep the file horizontal and file the tooth using two or three firm strokes. Now file each alternate tooth. Reverse the saw and file the other teeth.

Rip-saws The teeth are filed at right angles across the blade.

Cross-cut saws Hold the file at approximately 70° to the blade with the file handle pointing away from the saw handle (Fig. 4.23).

GRINDING DRILL BITS

The most important factor to take into consideration when sharpening a twist drill is that the two sides of the drill are equal. The angle of the drill point is 60°. When sharpening a drill, try and obtain one already ground to use as a guide.

1. Press one cutting edge of the drill lightly against the wheel at the required angle.
2. Turn the drill in a clockwise direction (as seen from above in Fig. 4.24) following the angle on the end of the bit (only grinding the one side of the drill).
3. Repeat on the other edge until both are the same.

TOOL BREAKAGES

It is unavoidable that tools will get broken. Don't jump to conclusions when a student breaks a tool as it may not be his or her fault – some tools are of notoriously poor quality. If the breakage was due to incorrect usage, be sure that you had instructed them to use the tool correctly before reprimanding them.

It is important that students are made aware of the importance of reporting broken tools and not be scared if they break one.

If a tool is beyond repair, remember to remove it from the inventory.

LIST OF COMMON TOOL BREAKAGES

The following faults were common in particular consignments of tools from Bulgaria and Czechoslovakia in use in Nigeria in 1989–90. They are given here as examples of the kind of faults you might have to deal with in the tools locally available to you. Particular attention should be given to safety precautions.

BULGARIAN EQUIPMENT

Smoothing planes Labelled as Jack planes but, in fact, smoothing planes. Two main faults have been found with these:

1. The cap iron is made from thin steel which allows shavings to enter between the cap iron and the cutting iron, which in turn clogs up the plane mouth. This means you have continually to dismantle the blade to remove the obstruction.
2. The frog is made from aluminium and, if the screws that hold the burr cap are over-tightened, they will strip the thread in the aluminium. If this happens, remove the frog and drill the hole right the way through and use a nut and bolt.

Ratchet brace Be very careful with these and perhaps limit them for demonstration. The jaws are made of a cast-iron and snap under the slightest pressure.

Claw hammers The claw of the hammer is not hardened at all. Consequently, the action of removing nails wears away the 'V' in the claw.

Screwdrivers Handles are prone to breakage. If possible, make wooden ones.

F cramps It may be found that these slip when tightened. If this is the case, drill 6mm holes at intervals along the length and use like a sash cramp.

Breast drill The handles are made from aluminium and break easily. New handles can be made from steel plate.

Guillotine The blade of this is fairly brittle because of its hardness. Only use on thin sheet and use it gently. It is a good idea to make a lock for this as it seems to be the student's favourite piece of equipment.

Bench vice The main nut that the handle runs through is made from very brittle cast iron. If the vice is over-tightened, it will shatter. Use with care.

CZECHOSLOVAKIAN EQUIPMENT

Hammers The shanks tend to come away from the head. It may be found that these can be fitted more securely.

Circular saw The switch is not very good and is potentially dangerous. Also the saw table is not very strong and vibrates badly when used.

File handles The plastic file handles crack. These should be replaced by wooden ones. If these cannot be made at the school, ask a local wood turner to make them.

Do not be disillusioned by the above. It is possible to find good solutions to most of the problems faced.

5. SCHEME OF WORK AND LESSON PLANS

A detailed scheme of work must be tailored to a particular school and a particular teacher but the same broad guidelines given in this chapter can be applied to any school or teacher.

CONTENT

The syllabus for introductory technology is very broad and, in most cases, to be covered reasonably must be reassessed and rationalized to suit each school's individual situation.

A problem many teachers have is knowing what to teach and in how much detail it should be taught. Obviously the scheme of work is there as a guideline but you will need to set your own criteria as to what should be taught.

The syllabus can be organized into 'must know', 'should know' and 'could know' subjects.

MUST-KNOW SUBJECTS

These are topics the student *must know* to be able to study introductory technology, e.g. measurement, materials, shapes, sketching.

SHOULD-KNOW SUBJECTS

Topics the students *should know* and which will help them to understand the many aspects of introductory technology.

COULD-KNOW SUBJECTS

Subjects not central to the aims and objectives of introductory technology and covered elsewhere in the curriculum or outside school.

It must be stressed that the choice of 'should know' and 'could know' subjects is subjective and does depend on each school's situation. For example, if the Agricultural Department is strong, related subjects in Introtech will be considered 'could know' subjects. Again this highlights the importance of working closely with other departments as already mentioned. To use these criteria the teacher must have an aim to work towards, e.g. is the 'must know' information essential for the passing of an exam or the basic understanding of the subject (these may be very different)?

TIME AVAILABLE

It is pointless intending to cover a lot of content in great detail, only to find that the lack of time available means having to rush, missing out content and detail haphazardly. It is far better to plan conservatively and have time left over to consolidate, to offer extra content or detail in an organized manner.

Due to the nature of school organization and requirements, teaching time is interrupted by activities such as grass cutting, meetings, marching, exams, rain, wage collection, etc. A good rule of thumb initially is to assume 30 per cent of your timetable time will not be available for teaching.

WRITING A SCHEME OF WORK

The first thing to find out is exactly what has to be taught, and below is an outline of how this can be done.

You will need the following:

- syllabus;
- past exam papers;
- knowledge of students' ability;
- areas of the syllabus covered by other departments.

Once these have been collected it is possible to cut the syllabus down by:

(a) looking at past exam papers to see which sections of the syllabus never come up;

(b) finding out from integrated science, agriculture and home economic teachers which areas of the syllabus they already cover. If it is found that certain sections of the syllabus are the same, then a decision will have to be made as to how much overlap there should be, if any;

(c) by giving an assessment test during the first lesson; from this you will be able to gauge what the students already know.

Once this has been done, you should have a more reasonable-looking syllabus. It may be found that, during the assessment test, student weaknesses surface that are not a large part of the syllabus, but are important so will have to be added to it.

Now the syllabus can be separated into 'must know', 'should know' and 'could know' subjects. From this a rough three-year scheme of work can be written (just listing the topics to be covered), then more detailed one-year and one-term schemes of work can be written, taking into account the timetable and time available.

ACTIVITIES WITHIN THE SCHEME

When writing a detailed scheme it is essential to know what activities are going to be used to teach particular subjects. These activities should be an integral part of the scheme and *not* something decided on afterwards. To do this, you must know what has to be covered in a particular lesson and which activities would help to teach the subject. For example, if you are teaching measurement, shapes and materials, activities such as making mobiles, jewellery and name-plates would be an effective method of teaching, allowing the student to learn by doing. This would not be the sole method of teaching the subject and would be supplemented by other activities, each one building on knowledge and skills gained from the other.

It is not possible to write a definite scheme of work at your first attempt. The scheme will need continual updating and initially only provides a framework in which to work.

SAMPLE SCHEME OF WORK

This sample scheme of work is written for three years, and details the basic subject matter that is to be covered each term.

NB: measurement, sketching, shapes, drawing, design, maintenance and safety should be incorporated into everything the student does. These are the basic life skills the student needs in order to be able to carry out other activities.

The teaching of tools and their uses should be 'hands on' with students learning by doing and from visual aids around the room. Time should not be spent specifically teaching about tools although short notes may be given explaining the finer points of a particular tool.

YEAR ONE

1ST TERM

Subjects What is technology? Measurement and units, sketching and labelling, drawing equipment, plane shapes (circles, triangles, squares and rectangles).

Activities Students to study examples of local technology such as farming tools, spirit distillation, soap making, etc. All measurement, sketching and drawing lessons to be pure activity lessons. Make mobiles, jewellery and toys. All the skills gained may be incorporated into a final project.

2ND TERM

Subjects Materials, properties and identification of metal, wood, ceramics, plastic and rubber.

Processing of materials (basic), production and forms of materials, uses of materials (basic).

Activities Make small clay items and fire them in a simple mound kiln. Experiments in the identification of materials and their properties; use materials used locally for experiments; small project using different materials with different properties, e.g. catapult (wood and rubber), puppet using bamboo and string, clay beads threaded on to wire, calabash carving; the metal carving tools show the different properties of materials very well.

3RD TERM

Subjects Maintenance, safety, simple machines, friction. Energy: forms, conversion, sources.

Activities
Maintaining bush lamps and bicycles, making a seesaw, simple pulleys, small toy using simple transmission (belt-driven). Discover by experiment sources and forms of energy.

YEAR TWO

1ST TERM

Subjects Technical drawing, geometrical construction of plane figures, basic electricity and magnetism, current, voltage, circuits, safety.

Activities Drawing practice, paper model making, magnetic experiments, making simple circuits, steady hand game (students to find bulb, battery and wire to make their own).

2ND TERM

Subjects Power, energy conversion based on common appliances, i.e., charcoal and electric irons, kerosene and gas cooker, solar and hydro power. Electro-magnets, batteries. Material-joining techniques and finishing (basic).

Activities Experiment with charcoal irons and kerosene stoves to discover more about energy conversion (chemical-to-heat energy). Experiment with coil water-heaters (electrical-to-heat energy). Use magnifying glass or similar to show solar-to-heat energy (burn a piece of paper). Teacher to design and make lots of models and toys to show different types of energy conversion. Using joining techniques students can make models showing energy conversion, e.g. very simple waterwheel or windmill. Make electro-magnets, strip down old dry-cell batteries, make battery using lemon.
NB: joining techniques and finishing should be used in all activities, so students learn from 'hands on' experience.

3RD TERM

Subjects Building, setting out and preparation, foundations, walls, doors and windows, floors, roofs, maintenance, sewage, sanitation and food preservation.

Activities If possible, a small building project may be started as an on-going project, e.g., latrine house, chicken house, storeroom, alterations to an old classroom or a pottery kiln. Bricks can be made from mud.

YEAR THREE

1ST TERM

Subjects Isometric drawing, orthographic drawing (1st & 3rd angle projection), water supply, purification, pumps, material processes continued.

Activities Drawing practice of projects made or to be made.

2ND TERM

Subjects Mechanics; linear, rotary motion, mechanical gears, belt drive. Electricity supply, household wiring, maintenance.

Activities Experiments using bicycles, sewing machines, pepper grinders and typewriters, to identify different forms of motion. Wiring plugs, maintenance of electric irons and coil water-heaters. The construction of toys or items that require the conversion of one type of motion to another.

3RD TERM

Subjects Woodwork and metalwork processes, forging, sheet-metalwork, screw threads, joining techniques and finishing continued, revision.

Activities Final third-year project combining the knowledge gained in the many areas of introductory technology.

SAMPLE LESSON PLAN: MEASUREMENT (1)

Time 2 × 40 min.

Visual aids Five 1m-long strips of card
Five one-metre rules
30cm rule

What? Today we are starting a new subject, 'Measurement' (on board). We will be looking at rulers and seeing how we can use them to measure things in metres and centimetres.

Why? Measuring is very useful in technology. We must learn how to measure before we go into the workshop so that if we want to cut a piece of wood, we can measure it first and make sure we cut it in the right place. If you want to go into the workshop for practicals, you must *first* learn to measure.

How? First of all I am going to divide you into five groups. Then each group will measure different things in different ways and I will keep all your results on the blackboard. We will also be finding out how to use a ruler.

Check pre-knowledge Call a number of students out to measure chalk lines of various lengths on the board. This should give a reasonable test of the standard of the class.

Development 1 Divide the class into five groups. Have one student from each group 'measure' the length of the classroom by counting how many paces from one end to the other. Note that different students get different 'measurements'. Repeat this with a different student from each group, but this time measure the length of the classroom in 'heel to toe' steps. Record all results on a table on the blackboard as shown:

	1	2	3	4	5
LENGTH OF CLASSROOM (paces)					
LENGTH OF CLASSROOM (heel to toe)					

Note that again different students get different results.

Q. If we were building a new classroom, and decided to make it ten paces long, and we let one student (name a tall student) measure one wall, and another student (name a short student) measure another wall, what would happen?

A. The walls would be different lengths.

Q. That is correct; the classroom would be like this: (show on board). What is a better way to measure out the classroom?

A. Use the same boy for all measurements.

Development 2 Give a 1m-long strip of card to each group. Have one member of each group measure the length of the classroom. Add the results to the table.
Note that each group has got the same answer.

Q. Why has each group got the same answer?
A. Because the strips of card are all the same length.
Q. If we were building a new classroom, and used these strips to measure it, what would happen?
A. The walls would be the same length.
Q. The classroom would be like this: (show on board). Is this better?
A. Yes.

'So we see that using these strips is better than paces.'

Development 3 Hold up metre rule. Show that it is the same length as the strips of card. 'This is called "metre rule" and is exactly one metre long.' Each group to use the metre rule to measure the length of the classroom.

Q. How many metres long is the classroom?
A. . . . . . . metres.

Enter results in the table, writing . . . . . . m.
Note that 'm' is used to show metres. Also note that the results are all the same.

Q. What is this called? (Hold up metre rule.)
A. Metre rule.
Q. How long is it?
A. One metre.

Development 4 Give each group a 1m strip of card.

Q. How long is this piece of card?
A. One metre.
Q. Use it to measure how wide your desks are. How wide are your desks?
A. Half a metre.
Q. Are they exactly half a metre?
A. No.
Q. Look at the metre rule. We can see that there are numbers all along it. What is the first number?
A. Zero.
Q. What is the last number?
A. One hundred.
Q. So the one metre has been divided up into 100 smaller pieces. We call these smaller pieces 'centimetres' (on board). Normally we write 'cm' (on board). (Give metre rules to each group.) Now

use your metre rules to measure how wide your desks are. Remember to put the zero on one edge of the desk. How wide are your desks?

A. . . . . . . cm.

Development 5 Use the remaining time to measure as much as possible in the classroom, e.g. length of blackboard, height of desks, charts, etc.

Summary Let us look at what we have done today. (Hold up one-metre rule.)

Q. What is this?
A. Metre rule.
Q. How long is it?
A. One metre.
Q. How many centimetres long is it?
A. One hundred centimetres.

'Next lesson we will carry on with measurement. We will be looking at how to measure small things.'

SAMPLE LESSON PLAN: MAINTENANCE

Time 2 × 40min.

Visual aids Chalkboard
Bush lamp
Bicycle
Post showing outline of a preventive maintenance schedule for a bush lamp.

What? Today we are going to find out about 'maintenance' (write on board).

Why? If we maintain things well, they will do their job better, last longer, and they will be safer. This means we can save *time, money* and even *lives.*

How? First I will explain about the different types of maintenance we can use and ask you some questions. Then, using the bush lamps you have brought, and my bicycle, we will practise maintenance.

Check pre-knowledge Ask students: What is maintenance?

Development 1 There are two types of maintenance; preventive maintenance (write on board); corrective maintenance (write on board). Preventive maintenance is when we do something to stop the item of equipment from going wrong or spoiling.

Corrective maintenance is when we have to repair the item of equipment in order to make it work again.

Q. If your water bucket has a hole in it, what type of maintenance will you have to carry out to make it good again?
A. Corrective maintenance.
Q. Do you think it's better to carry out preventive maintenance or to wait until something spoils and carry out corrective maintenance?
A. Carry out preventive maintenance.
Q. Give one example of preventive maintenance.
A. Oiling a bicycle chain.

Development 2 When we want to carry out any type of maintenance it's a good idea to write a 'maintenance schedule'. This tells us what we need to do and when to do it.
Let's look at the bicycle (put it on the table) and see what is wrong with it.

Q. What is wrong with the bicycle?
A. The front tyre is flat, the chain has fallen off and the seat is loose.
Q. What type of maintenance will we have to carry out?
A. Corrective maintenance.

'Let's make a corrective maintenance schedule for the bicycle.' (Draw schedule outline on the board and fill in using Q & A.)

Corrective Maintenance Schedule for the Bicycle

What's wrong	Corrective Maintenance
Flat tyre.	Pump up. If this doesn't work, fit a new inner tube.
Chain has fallen off.	Put it back on and oil it.
Seat is loose.	Tighten the bolts up.

Development 3 (Show poster) I now want you to write a preventive maintenance schedule for your bush lamp. (Give full instructions to students on how to carry out the activity. Go around the class giving assistance where necessary.)

Development 4 Preventive Maintenance Schedule for a Bush Lamp

What we need to do	When we need to do it
○ Wash the glass	Every week
○ Wash the body	Every three months
○ Wash roller mechanism	Every three months

○ Trim wick	Every week
○ Decoke top	Every week

When the students have finished their work, go through it on the board using Q & A.

Summary Let's look back at what we have done today.

Q. Name the two types of maintenance?
A. Preventive and corrective.
Q. What is a maintenance schedule?
A. It tells us what to do and when.
Q. How can we reduce the amount of corrective maintenance we need to carry out?
A. Carry out preventive maintenance.

'Now that you know about maintenance you can prepare your own maintenance schedules for other items in your house, such as your stove or farming tools. It may be a good idea for some of you to carry out corrective maintenance on your exercise books!'

SAMPLE LESSON PLAN: SKETCHING

Time 2 × 40min.

Visual aids Resistor
Assortment of tools, household items, etc.
Sketch of a resistor
Assortment of good clear sketches.

What? Today we are going to find out about sketching.

Why? Whenever we want to make something we need to know what we are going to make and what it will look like when it is finished. A sketch will help us to understand these things.

How? I will give you an example of the importance of sketching. Then you will all practise sketching certain objects.

Check pre-knowledge

Development 1 Ask for a boy and a girl to come up to the front of the class. Show them and the whole class an assortment of items including a resistor. Ask both of them to find the resistor giving only one of them the sketch.

Q. Why do you think the person with the sketch found the resistor first?

A. Because they could see what a resistor looked like.

'Now we can see the importance of sketching. If we cannot describe what we want through speech, we can use the language of sketching/drawing.'

Development 2 When we sketch we use a pencil. This is so we can erase parts of the sketch if we change our minds or make a mistake. There are two other things we must know before we start practising sketching. (Show example of a large clear sketch.)

Q. Can you see this sketch?
A. Yes.
Q. Why can you see it?
A. Because it is large and clear.
Q. Why does it have to be large and clear?
A. So other people can understand it.

'Good. So now we know that sketches have to be large and clear so people can understand them.'

Development 3 Now it is your turn to sketch. I will give out the items you are going to sketch (saws, buckets and cooking tripods). (Give full instructions on how to carry out the activity. Give help to students on a one-to-one basis.)

Summary (Open praise for students' work)

Q. Name two things we must do to make our sketches easy to understand.
A. Make them large and clear.
Q. Why do we need to make sketches?
A. To help us understand what something looks like.
Q. (Hold up a sketch of a farming hoe) What is this?
A. Farming hoe.
Q. Using this sketch do you think we can make a hoe like this?
A. Yes.
Q. How big should we make it?
A. Don't know.
Q. How can we find out how big to make it?
A. Measure an old hoe.
Q. What should we do with the measurements.
A. Write them on the sketch.

'Although the sketches you have done today are very useful, we can make them even better by writing the measurements on them. In the next lesson we will measure the things you have sketched today and then put the measurements on the sketches.'

SAMPLE LESSON PLAN: BASIC ELECTRICITY

Time 40 min.

Visual aids Large circuit-board with movable components which will be built up by students.
Small circuit-boards to be given to groups
Worksheet for experiments to be written on board.

What? Today we are going to find out how electricity lights an electric bulb and what materials make good conductors.

Why? It's very important that we know how electricity works, as it helps us in our everyday lives.

How? I will explain a few important things we need to know about electricity, then you will carry out an experiment using a worksheet which will be written on the board.

Check pre-knowledge

Development 1 To make a bulb light (show bulb) electricity has to pass through it. The electricity we are going to use is stored in a 'dry-cell battery' (show). So the electricity can flow from the battery to the bulb we have to attach wires (show wires) from the battery to the bulb. Electricity flowing in a wire is called electric current and is measured in 'amps'. For electric current to flow there must be a circuit. A circuit is a complete path around which the electric current can flow.

Q. What metal is our electrical wire made from?
A. Copper.
Q. Why is copper used?
A. Because it is a good conductor of electricity.

'Good. You have used the information you learnt in a previous material lesson to solve this problem. Once again this shows us that all subjects are connected together.'
(Using a large circuit-board at the front of the class, start to build it up using Q & A.)

Q. What is the first item we need?
A. A dry-cell battery (student to put on board).
Q. What else do we need?
A. Bulb (student to put on the board).
Q. What is the final item we need?
A. Wire.
Q. How can we make the bulb light?
A. Fasten the wires from the bulb to the battery (teacher to attach).

'Now we can see that our bulb is shining. Because the bulb is shining, it proves that electricity is flowing from the battery to the bulb.'

Q. What do we call this?
A. Circuit.
Q. What do we call it when electricity flows through a wire?
A. Electrical current.

Development 2 We have been told that electricity will only flow through a conductor, with copper being the best. Materials that do not conduct electricity are called insulators.
Now by carrying out experiments we are going to find out which materials make good conductors and which make good insulators.
After giving full instructions, students are to carry out experiments from tho worksheet.

Development 3

Experiment Small circuit-boards containing a battery, bulb and wire to be given out. Students are to test the conductivity of materials by connecting the ends of the wires to the material being tested and seeing if the bulb lights.
Students to draw a chart and fill in.

Conductors	Insulators
Steel spoon	Plastic biro
Brass key	Stone
Aluminium coat-hanger	Glass
Pencil lead	Wood

When students have finished their experiments, fill in the chart on the chalkboard using Q & A.

Summary
Q. Where did the electricity come from that we used for this experiment?
A. Dry-cell battery.
Q. How did the electricity get to the bulb?
A. Through the wire.
Q. What do we call electricity when it flows in the wire?
A. Electric current.
Q. What is an insulator?
A. A material that does not conduct electricity.
Q. What could we use an insulator for?
A. To stop electricity flowing somewhere.

Q. Give an example of an insulator?
A. Plastic on a wire.

(Light the bulb on the circuit-board.)

Q. What is it called when the electricity flows along a complete path?
A. A circuit.

'In the next lesson we are going to start making an electrical game which you can take home. During the next week, try and find any dry-cell batteries that still work, and torch bulbs.'

6. CLASSROOM MANAGEMENT

This is possibly the most important chapter in this book. Even if you have access to a well-equipped workroom, unlimited supplies of materials and a perfect timetable, the effectiveness of your teaching depends ultimately on your classroom management.
Good classroom management means:

1. Keeping good records.
2. Gaining control and keeping the respect of your students.
3. Getting to know your students.

RECORD-KEEPING

Good record-keeping is the key to an organized classroom. You must always know who was late; when and why; who has and hasn't handed in their assignments; what work has been covered and how well the students understood the work, etc.
Therefore it's very important the teacher keeps a register at the beginning of each lesson which will soon identify the students who are continually late, not giving in assignments, etc.
The teacher should also keep a personal diary detailing each lesson taught, noting the following:

1. Work covered.
2. The effectiveness of the lesson.
3. Whether or not the students were well behaved and on time (and if not, why).
4. Work given out to students and date expected in.

The recording of marks is covered in Chapter 8: Assessment.

GAINING CONTROL AND RESPECT OF YOUR STUDENTS

A vital key to good classroom management is classroom control. The essence of this control centres around the relationship you form with the students. In helping to create a good student/teacher relationship you should:

ALWAYS DRESS SMARTLY

There is no excuse for being untidy.

GET TO KNOW YOUR STUDENTS

It is essential that you learn the names of your students. At first the students may laugh when you pronounce their names wrongly: you should take the joke and try again. *Never* get embarrassed by wrong pronunciation of names. Here are some ideas on getting to know students' names:

- Take a register at the beginning of each lesson.
- Walk around the class when students are working and look at the names on their exercise books.
- Hand students' work back yourself.
- Use a seating plan.
- If you cannot remember a name, ask.

PRAISE YOUR STUDENTS

Students will be much happier in their work and more respectful of the teacher if praise is used constantly, group praise as well as individual praise. A quiet 'well done' in a student's ear as you walk round is good for self-confidence.

SET RULES OF THE CLASSROOM

Students have to be told how you expect them to behave in class, therefore rules have to be set. These rules should be given to the students in their very first lesson (it is a good idea if they write the rules in their books). They must also be aware of the consequences of breaking the rules. The most important point concerning rules is to *stick to them and be consistent*. There should never be exceptions.

Once it is known a teacher means what he/she says, the teacher will be respected by the students.

Here is a sample set of rules for the students:

- Always arrive on time.
- Always bring biro, pencil and exercise books.
- Always be quiet when the teacher is talking.
- Always put your hand up to answer a question.
- Always hand assignments in on time.
- Always ask if you don't understand.

Where possible, rules should be expressed positively since a positive statement offers a goal to work towards rather than something to avoid.

GIVE CLEAR INSTRUCTIONS

Students must received clear instructions for every assignment or activity they carry out.

STARTING THE LESSON

It is essential that the lesson starts in an orderly fashion as it sets the tone for the rest of the period. Here are a few basic tips for getting off to a good start:

- Be at the workroom on time.
- Control the pupils' entry into the classroom.
- Get total silence before speaking to the class.
- Start the lesson with a 'why' which will sustain interest and curiosity.

The best method of getting silence is to stand in front of the class in silence yourself. The class will rapidly quieten.

DURING THE LESSON

Once the standard of classroom management has been set, it must be maintained throughout the lesson. The teacher should:

- Learn and use students' names.
- Give clear instructions.
- Give praise.
- Be mobile, walk around the class.
- Look at the whole class when talking (by scanning) and involve *all* students.
- Anticipate discipline problems and act quickly.

FINISHING THE LESSON

Finishing the lesson in an orderly fashion is as important as starting it so. You must know exactly what your are going to say and do. If tools, drawing instruments, books, etc. have been handed out, these must be collected in without causing a disturbance. The teacher should:

- Be aware of the amount of time needed for clearing up and summing up.
- Have a system for collecting in work, instruments, etc. at the end of the lesson. (Make sure you and your students know the system.)
- Get silence before the students leave.
- Remind students of any items needed for the next lesson.
- Control students when leaving the classroom.
- Always finish the lesson on time to avoid students being late for their next lesson.

DISCIPLINE

By being both firm and consistent, you should not encounter many discipline problems. However, there will always be times when a student has to be reprimanded. The timing of a reprimand is critical.

It is easier to stop a careless whisper than a full-scale riot! If a reprimand is necessary, then act quickly and above all act fairly. Move potential troublemakers to the front of the class where they can be easily watched. When reprimanding a student, remember the following points:

- Never make idle threats that you will find hard to enforce.
- Don't be hasty when handing out punishments. What at first seems like an act of indiscipline could be a reflection of a genuine problem. For example, a student who seems to be sleeping may be ill.
- Never give punishment because of personal feelings.
- Never punish a student by giving extra schoolwork, i.e., Introtech.
- Wherever possible, the punishment should be corrective. For example, if a student has written on a desk, he/she should clean it.

Whatever happens, you will always remain an adult and a teacher; and your charges will always remain young and pupils. This is how it should be – do not upset the balance by trying to change this. It is all part of the professional relationship that is the mark of a successful teacher.

Fig, 6.1 Safety in the workroom

SAFETY IN THE WORKROOM

The majority of students' time will be spent carrying out practical activities, which means using tools – and using them safely.

With so many students using tools at the same time, it is essential that the students' and teacher's safety awareness is high. As long as students are well instructed in tool use there should be few problems.

At first the teacher may be shocked at what seems like very poor safety standards, e.g. students wearing flip-flops/bathroom slippers in the workroom. However, the fact of the matter is that these may be all the students have. The teacher has two choices, either:

1. Do not have any practical lessons; or
2. Be aware of the situation and make students aware of the dangers involved.

One way of creating good workroom control, which helps to avoid situations which could lead to accidents, is to teach all the students to stop work *immediately* you say 'Stop'. You may be able to see a potential accident at the other end of the workroom and, if all students stop on your command, the accident could be avoided.

Obviously, the teacher should not flaunt safety rules but rather reassess certain situations. For example, it is obvious that safety glasses must *always* be worn when grinding.

Fig. 6.2 Class size

CLASS SIZE

The number of students in each class varies enormously from school to school, but many teachers are teaching forty or more students per class. Don't despair! Activity work can still be carried out effectively. However, the teacher must be very well organized. (The importance of not having an overloaded timetable can be seen.) The following are a few methods that have been used successfully:

GROUP WORK

Putting students into activity groups is an ideal way of teaching introductory technology. Nigerian children are used to working together, helping each other and sharing (although the teacher must still check all students are getting practice in certain areas).
The number of students in each group will vary according to class numbers but between four and six is fine. Here is an example of a group lesson:

Objectives

1. Students to be able to identify the following materials:
   metal (ferrous + non-ferrous)
   wood
   plastic (thermo-set + thermo-plastic)
   ceramic
   rubber
2. Students to discover the different physical properties of the materials by experiment.

Activities Students to split into their groups and carry out experiments with the aid of a worksheet.

Notes Each group will be given samples of each material and a worksheet explaining how they can carry out experiments.
If the activity has been well planned and is interesting, the students will soon become totally engrossed with their projects.
Teaching forty students by this method is not a problem if the teacher has created enthusiasm and has organized the lesson in such a way that the students are eager to discover the solutions/answers to the problems set.

SPLITTING THE CLASS

Another method of coping with large classes is to actually split the class into two. Each lesson, half the students will carry out activities while the other half do drawings, notes, worksheets, etc. Depending on the style of the workroom, students can either stay in the workroom, with half doing practical and half drawing etc., or if this is not convenient, half can go to their class to work.

Logistics of splitting the class First, both teacher and student must know exactly who is in each class/group. Then a rota must be made for the groups. This can take considerable organization.

If at all possible, it is best to keep the class as one on various activities in the workroom and outside the workroom. By splitting the class there is a danger of creating the theory-and-practical situation which the teacher should try and avoid.

NB: do not forget all the space outside the workroom. Many activities can be done outside.

EVENING/AFTERNOON CLASSES

Many teachers find that because of time constraints, extra afternoon/evening classes have to be organized. Before these classes are planned, the following must be taken into consideration:

- If it is a day school, will the students be able to come back?
- If it is a boarding school, what duties do the students have to carry out after school?
- What other extracurricular activities take place and on what days and at which times?

Once the teacher has found out the above, they can then discuss the proposed lessons with the Principal and any teacher that could be affected by the classes.

Do not make the evening classes too long. These are usually very enjoyable lessons for both teacher and student, and the teacher may be tempted to extend the lesson. It will be counter-productive in the end, as it will be accepted as the norm, the teacher will become overworked, and often students lose interest.

7. TEACHING METHODOLOGY

This chapter gives examples of how the teacher can actually get information over to the students. It explains about visual aids, teaching aids, worksheets, giving notes and setting up a resource section in the workroom.

TEACHING AIDS

A teaching aid is anything that helps a teacher to teach and a student to learn. Outlined below are some examples of teaching aids which, if used separately, or together, form a very efficient method of teaching.

WORKSHEETS

The use of worksheets enables the teacher to teach mixed-ability classes, whilst the students are able to work at their own pace and discover things for themselves.

CONTENTS OF WORKSHEETS

1. Instructions: these should be clear and precise, explaining how to use the worksheet, and its aims.
2. Information: it should be written/drawn in a simple, clear and attractive manner, stimulating a desire to learn. The more visual it is, the better.
3. Assignment/activity: based on information given, the assignment or activity should allow for discovery learning. Remember, 'It's fun finding out'.

The following are two methods of setting worksheet assignments:

(a) Graded assignments: this is when questions are set at different levels. The first five may include all the 'must know' information and have to be answered by all students, whilst the other five questions are optional and include the 'should know' and 'could know' information.

(b) Fixed-aim worksheets: these are targeted at the most able students, who work unsupported, whilst other students work with the support of the teacher or further information sheets.

4. Dictionary: all new words used can be listed at the end of the

worksheet for the students to copy into their books. This way they will soon build up a considerable technical dictionary.

Although worksheets do take considerable preparation, once done they provide a permanent resource for both teachers and students. Students do have 'fun finding out' and the teacher will gain satisfaction from being able to work closely with students.

NB, the word 'worksheet' is used to describe a method of teaching rather than a physical printed sheet. Although this method suits printed sheets it is just as effective when the contents are written on a blackboard or large sheet of card.

SAMPLE WORKSHEET

MAGNETS AND MAGNETIC FIELD

INSTRUCTIONS

Read the information carefully. Then carry out the experiment. Write all answers in your book.

Remember If you do not understand, ask.

INFORMATION

A magnet is a metal object which attracts certain kinds of metals. The metals the magnet attracts are called 'ferrous metals' and contain iron. The metals that are not attracted to the magnet are called non-ferrous metals and do not contain iron.

A magnet has a 'North pole' N and a 'South pole' S.

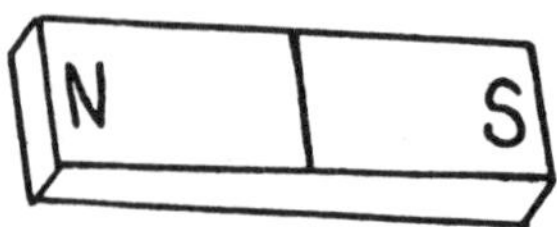

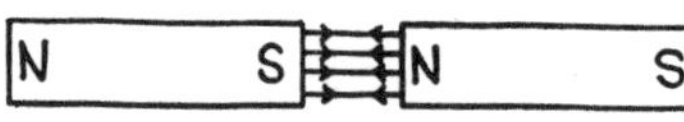

When two magnets are put together with opposite poles, they will *attract* each other.

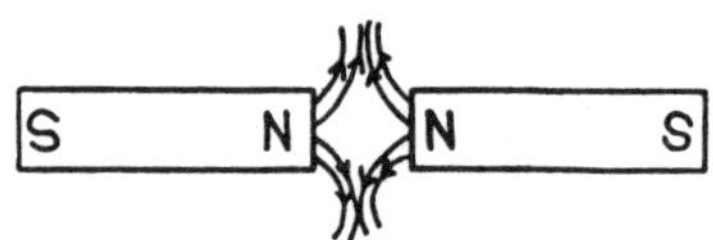

When two magnets have similar poles, they will *repel* each other.

This shows that magnets create a force. We call this the 'magnetic field'.

Have you read the information?

EXPERIMENT

On your table you will find the following: two bar magnets;
piece of paper;
iron filings;
pieces of metal with numbers on.

Let's find out more about magnets.
Using the magnet, try and pick up the pieces of metal.

Q.1. Which numbers did the magnet pick up?
Q.2. Is this metal ferrous or non-ferrous?

Using the two magnets, try and make them stick together.

Q.3. What are the names of the two ends that stick together? Are they:
(a) North and South?
(b) South and South?
(c) North and North?

Place the piece of paper *on top* of your magnet.
Very carefully sprinkle the iron filings on to the paper. Tap the paper gently.
This shows the 'magnetic field'.

Q.4. Draw this in your book.
Q.5. Repeat the experiment with two magnets as shown overleaf.

N S

N S

Draw the pattern in your book.

Q.6. Repeat the experiment once more with the magnets positioned as shown below.

Q.7. Look at the drawing below. Will the magnets *attract* or *repel* each other?

Q.8. Is the name of the magnet you used: (a) Horseshoe magnet?
(b) Beam magnet?
(c) Electro-magnet?
(d) Bar magnet?

DICTIONARY

Write these words and definitions in your dictionary:

Ferrous metal: a metal that contains iron and is magnetic.

Non-ferrous metal: a metal that does not contain iron and is not magnetic.

Magnetic field: space around a magnet in which a force is exerted.

Magnet: find this out yourself.

Attract: find this out for yourself.

Repel: find this out for yourself.

When you have finished put all the tools and materials back in the box. Well done.

QUESTION BANKS

A question bank is simply a store for a large amount of questions. Large question banks can be built up using worksheets, pictures and textbooks as the information source.

If textbooks are used, questions can be set from specific pages, e.g. students who finish their work before anyone else can be given an appropriate textbook (from the resource area) and a card from the question bank. The chosen card will ask questions taken from information on a particular page of the book.

TECHNICAL LIBRARY

All books and pictures relating to introductory technology can be built up into a library. Pictures with a sentence describing the item shown make a simple reference system.

VISUAL AIDS

Visual aids are a must for the workroom. They help to explain things quickly and simply and can be used again and again. The basic rule to follow when using visual aids is to make them:

Simple – to understand
Bold – easy to see
Exciting – interesting to students
Fun – for teachers and students

Examples of visual aids:

Chalkboard
Posters
Flash cards (cards with drawings or word on)
Models; papier mâché is one of the best methods of making models (see Chapter 12)
Working models
Mobiles
Flannel board: easily made by using an old blanket as the flannel; sandpaper may be stuck to the back of visual aids which then stick to the blanket.

To make visual aids simple, bold and exciting, use the following:

Pictures Snaps, prints, drawings and cartoons.
Lettering Use different styles (make sure it can be seen from the back).
Colour Use contrasting colours or a black outline. (If there is a shortage of marker pens, colour your posters with paint.)
Materials Use available materials (if you don't have a sheet of card, it doesn't mean you can't have a poster!). Don't forget collage – materials can be fastened to old mats to make excellent 'posters'.

PROTECTING POSTERS

Posters can be covered with clear plastic – if available. An alternative is to coat the poster with some type of clear liquid that will set. One method of doing this is to water down white PVA glue; this can then be brushed on to the poster. Experiment on card first until you get the water/glue mixture correct.

If putting posters on the wall, remember to change them or move them around regularly, keeping the workshop interesting for the students.

NB: be very careful if posters show section views. Sections are a difficult concept for many students.

IDEAS BOOK

A must for any teacher. Find the biggest exercise book you can and write down any activity ideas, material information and any brain-waves you may have.

GIVING WORK AND NOTES TO STUDENTS

Because many students will not have textbooks, they will have to do work in their exercise books, thus creating a 'mini' textbook of their own. Here are a few ideas on how students can effectively produce work in their books. Remember to:

(a) explain to students how to lay out their work;
(b) make sure your own writing and drawings are neat;
(c) be aware that drawings and writing may take a long time – do not set too much;
(d) make notes short and punchy;
(e) test that students understand what they are writing and drawing;
(f) be very careful if work is being copied from textbooks – is it all relevant? Are the drawings good?
(g) find methods of giving notes in an interesting manner, i.e., missing words;
(h) decide if notes are to be given at the beginning, during or end of the lesson;
(i) beware of analogies; they may be taken literally.

COPYING WORK FOR STUDENTS

A lot of towns have photocopying machines or carbon duplicators. For important work such as sample question sheets the students may be willing to pay for a copy. If no copying machines are available, a jelly duplicator may be used (see Chapter 12).

Do not forget – introductory technology is an activity subject where students 'learn by doing'. Avoid giving long, laborious notes.

LANGUAGE

English may be the second or third language for many students, and they may therefore have little opportunity to practise and hear the language necessary for reinforcement. Access to material such as books and magazines may be limited, thus impairing reading ability.

For the introductory technology teacher this lack of adequate English may be a barrier to communication and hinder learning for the student. However, there are a number of strategies that can be implemented which will aid the student and teacher.

DEMONSTRATIONS

Wherever possible, there should be demonstrations, so if the language used in the lesson is difficult the student will learn through seeing and doing.

SPEECH

Speak slowly and clearly using short sentence construction.

QUESTIONING

Check learning by questioning and reinforce the answers using language to broaden knowledge.

VISUAL AIDS

If the spoken or written explanation is a problem, then the old saying 'A picture says it all' comes to the fore. Visual material is generally more easily recalled than verbal. To check on the learning taking place, make the visual aids into games where the students are involved – again 'doing' as opposed to just looking.
Language need not be a great problem if lessons are visual and exciting. The key to the language problem can be summed up:

I hear – I forget; I see – I remember; I do – I understand.

MIXED ABILITY

Mixed-ability classes are a problem faced by all teachers, Listed here are a few methods that may be used to alleviate the problem:

WORKING AIDED/UNAIDED

If work is being given to students to carry out in class, they can all work at their own pace while the teacher gives extra support where needed.

EXTRA CLASSES

These can be arranged for students who need extra assistance. It is most important that these classes are seen as an advantage, not as a punishment. The classes should be fun and contain the three core skills: spacial awareness, measurement and language.

GROUPING – MIXED GROUPS

Students may be put into small mixed-ability groups where the quicker students assist the slower ones.

STREAMED GROUPS

Alternatively teachers can organize groups according to the individual ability of the student, thus allowing the teacher more time with the slower students. Shown here is one method of streaming students:

(a) There should be three groups divided according to the criteria:

'must know'
'should know'
'could know'

This means that all groups will cover the 'must know', two groups will cover the 'should know' and one group will cover the 'could know'.

An identity should be given to each group by the teacher, avoiding elitist-type classification (e.g. A, B, C or 1, 2, 3). Neutral names such as Reds, Blues and Greens should be used.

(b) It is important that students' progress is continually monitored to ensure they are in the appropriate groups. It is essential that the assessment is objective and unaided.

Although problems will always exist within a mixed-ability class, by getting to know the students and their strengths and weaknesses, the effectiveness of the teacher will improve greatly.

8. ASSESSMENT

This chapter is split into three sections: the first explains the continuous assessment system which is used by JSS schools, the second discusses the interpretation of students' marks and marking schemes, whilst the third shows methods of assessment to establish new students' ability.

JSS CONTINUOUS ASSESSMENT

In Nigerian schools all assessment is continuous, i.e. students are regularly tested throughout their school life. The marks of these tests go towards their school-leaving examination. For JSS students, 60 per cent of their final marks come from continuous assessment whilst 40 per cent come from the final exam.

HOW THE SYSTEM WORKS

Each term students will take *three* tests and *one* exam. All are set by the subject teacher except for the JSS final exam which is set by the Ministry of Education.

The format of the tests varies enormously from school to school. In some schools 'test weeks' are set aside while in others the teachers give tests at their own discretion. The important factor is that at the end of each term the student has three test marks which represent 60 per cent of their marks. At the end of each year the student will have marks from 12 tests and three exams. From these the following happens:

JSS YEAR ONE:	10% of all marks are calculated.
JSS YEAR TWO:	20% of all marks are calculated.
JSS YEAR THREE:	30% of all marks are calculated.

This gives the 60 per cent continuous assessment which is added to the third-year external exam worth 40 per cent.

REPORT CARDS

At the end of each term the teacher has to fill in student report cards. Again the format of these varies considerably. The teacher should be aware of the particular method used at the school, otherwise they may find themselves with a lot of extra work at the end of the term.

INTERPRETING ASSESSMENT

Once a teacher has marked a set of test papers they should then carry out a personal assessment of the student's performance. It is not enough to know that Grace got 60 per cent and Jacob got 23 per cent. Why did Jacob get 23 per cent? What mark did he get last time? Has he done better this time?

By looking carefully at each student's marks the teacher can soon build up a picture of each student's progress, thus discovering who needs extra help and in what area.

Personal assessment of each test/exam also allows the teacher to find out if *they* are teaching effectively and setting tests/exams at the correct level. If, for example, when marking an exam it is found that 90 per cent of the students get Question 10 wrong, then the teacher

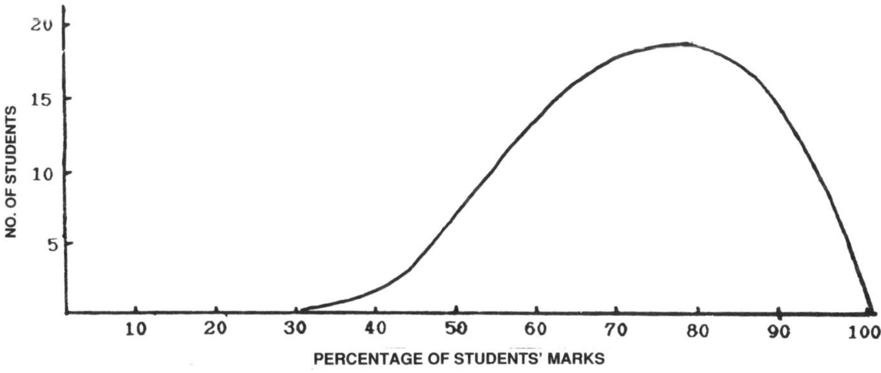

THIS DISTRIBUTION CURVE SHOWS THAT THE MAJORITY OF THE STUDENTS GAINED VERY HIGH MARKS. (WAS THE TEST TOO EASY?)

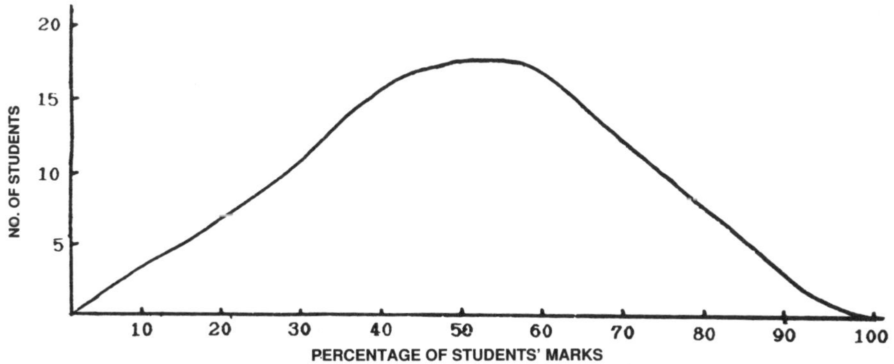

THIS DISTRIBUTION CURVE IS THE STANDARD SHAPE THAT SHOULD OCCUR FROM A WELL-SET TEST WITH A MIXED-ABILITY CLASS.

Fig. 8.1 Distribution curves

should look back to the appropriate lesson as it would seem that 90 per cent of the students did not fully understand the lesson.

DISTRIBUTION CURVE

A useful method of discovering if tests are set at the correct level is to construct a graph to show the distribution of marks as shown in Fig. 8.1.

MARKING SCHEMES

All assessment should have a marking scheme whether it is a ten-minute test, practical activity or a two-hour exam. The marking scheme should list the *question/answer(s)* and *possible marks*.

If a marking scheme is made as objective as possible – i.e., questions are either right or wrong – it will be easy to use, give fairly standard results and will be transferable to other teachers.

PRACTICAL ASSESSMENT MARKING SCHEMES

Owing to the 'learn by doing' nature of introductory technology something more than theoretical assessment is necessary. The following are two factors to take into account when making a marking scheme for an activity or project:

(a) Break the activity/project down into components, isolating the areas that the teacher wants to test.

(b) Each component of the marking scheme must be correctly weighted. (This applies to any marking scheme.) For example, if the teacher primarily wants to test measurement skills through a project, then the emphasis of marks should be for measurement rather than finish.

Below is a marking scheme for a pencil/pen holder. The main areas the teacher wanted to assess were measurement and angles, but they also wanted to introduce design skills that would develop individualism in the student.

Activity: Student's name: Class:

Component	Poss. mark	Act. mark	Comment
Dimensions			
Height 70mm	2		
Pos. of hole 1	2		
Pos. of hole 2	2		
Pos. of hole 3	2		
Angles			
30° angle	2		
Design			
Personal design	1		
Overall finish	1		
TOTAL	12		

NB, if the students are new to using tools, the teacher may mark the measurement before they cut their work. This way the teacher can make an accurate assessment of students' measuring abilities and not tool-handling skills.

It's often a good idea to show students the marking scheme before they start. They will then know which parts of the project are important and what is expected of them. If using worksheets, the marking scheme may be written on the back.

A personal assessment form can be produced and given to the student, or if there is a paper shortage, it can be written in the back of the student's book.

RECORDING MARKS

Once the teacher has marked students' work it's essential that the marks are recorded accurately. To do this the teacher *must* have a personal mark book. This will allow the teacher to:

- record students' marks efficiently;
- follow students' progress;
- assess their own teaching;
- fill in school documents such as reports and leaving certificates;
- run an organized class, knowing who has done what and when.

Fig. 8.2 is an example of a useful format for a personal mark book.

	CLASS 3B	LESSON 1 ASSIGNMENT	MOBILE PROJECT	LESSON 3 ASSIGNMENT	MONTHLY TEST W	MONTHLY TEST P
1.	SUNDAY IME	4	P	P	13	12
2.	TIJANI BASHIR	9	12	7	8	14
3.	STEVEN EKPO	AS	2	A	6	9
4.	MOHD MUSA	16	L 6	12	9	14
5.	MICHAEL OKON	P	17	AS 11	16	S
	DATE	15/1	22/1	29/1	2/2	4/2

KEY
W – Written
P – Practical

KEY
L – Work late
A – Absent when work was due
AS – Absent when work was set
P – Absent with permission
S – Sick

Fig. 8.2 Sample mark book

ASSESSING NEW STUDENTS

The first thing a teacher should do when meeting a new class is to assess each student's ability. If the class has never been taught introductory technology before, the teacher can test the basic knowledge they should have gained at primary school.

A short test can be designed to cover the following areas: mathematics, English, sketching, and spatial awareness. From the results the teacher will be able to assess the level of each student and then design a scheme of work and method of teaching suitable to the students' needs.

SAMPLE TEST
INTRODUCTORY TECHNOLOGY ASSESSMENT TEST

INSTRUCTIONS: Complete the following information.

NAME: .... AGE: ....

NAME OF YOUR VILLAGE: ....

FATHER'S NAME:......... MOTHER'S NAME:.........

FATHER'S OCCUPATION:...............................

HOW MANY BROTHERS DO YOU HAVE?.................

HOW MANY SISTERS DO YOU HAVE? ....

MATHS

FIND THE ANSWER TO THE FOLLOWING QUESTIONS:

Q. 1. (a) $2 + 4 =$ (b) $6 + 5 =$ (c) $12 + 9 =$

Q. 2. (a) $4 - 3 =$ (b) $9 - 2 =$ (c) $14 - 9 =$

Q. 3. (a) $2 \times 2 =$ (b) $3 \times 4 =$ (c) $4 \times 5 =$

Q. 4. WRITE ALL THE NUMBERS BETWEEN 1 and 20.

Q. 5. DRAW A SQUARE.

Q. 6. DRAW A TRIANGLE.

Q. 7. DRAW A CIRCLE.

Q. 8. DRAW A STRAIGHT LINE.

Q. 9. DRAW THE PICTURE ON THE BLACKBOARD.

Q. 10. USING COMPLETE SENTENCES, WRITE DOWN THE THINGS YOU LIKE BEST ABOUT SCHOOL.

NOTES

The information at the beginning of the test is designed to give the teacher individual knowledge of each student and to test the students' writing ability.

The drawing (Q. 8.) may be a simple isometric sketch of a box.

The aim of the last question is to test students' ability at sentence formation.

To test spatial awareness further, the teacher may organize a simple test that may be carried out while the students are working. For example, students can be called up to the desk one at a time and asked to fit a triangular shape into a matching hole; a ten-second time limit may be set for the successful completion of the task.

MARKING STUDENTS' BOOKS

Marking students' books can be a real chore but is an integral part of teaching. By marking books, the teacher will see how individuals are progressing. Giving coloured stars or badges out for good progress is a good motivator (which doesn't necessarily mean highest marks).

Fig. 8.3 Marking students' books

SETTING ASSIGNMENTS/HOMEWORK

Personal assessment of students' progress may be done through setting assignments. Remember that an assignment is as important as any test or exam, so the same amount of preparation and care must be taken, including a full marking scheme.

TYPES OF ASSESSMENT

Methods of assessing students' progress take many different forms, such as verbal questioning, quizzes, word association games, practical activities, written tests and exams.

Because of the students' educational background and the type of exams written by ministries it is best to assess as objectively as possible. It is also important to remember what is being assessed – that is, the students' knowledge of introductory technology.

The following are two examples of how a question may be set (which again highlight the problem of the language barrier):

Q. 1. A material which allows easy passage of electric current is called:

(a) transporter?
(b) resistor?
(c) insulator?
(d) contractor?
(e) conductor?

Q. 2. Explain what a conductor is.

Both these questions are asking the same thing. However, students with a low standard of English may not be able to answer Question 2 although they may know the answer.

The JSS external exam set by the Ministry is mostly multiple choice-type questions. Therefore by giving these type of questions during your tests students will become familiar with the methods of answering.

Special note: If certain students have continually poor results, the teacher should find out if the student has a special problem, e.g. deafness, poor eyesight, dyslexia or learning difficulties.

9. SKILL TRANSFER

It is one thing setting up a good system for you to work by, while it is quite another to have a system that other teachers can use and follow. This chapter gives a few ideas of how, once established, introductory technology can continue to be taught to the same standard through other teachers and workshop assistants.

THE INTROTECH DEPARTMENT

The size of the department will vary from one to maybe six teachers in each school. If there are several teachers in a department it is important that they work together from the start. The following are some ideas of how a department can work efficiently:

- It is absolutely essential that the whole department understands and is willing to work towards the aims of Introtech (Chapter 1).
- A head of department must be appointed.
- Regular meetings should be held to discuss progress within the department.
- When a scheme of work or timetable is being written the whole department must be involved.
- The Principal must be kept informed of what is happening within the department.

COUNTERPART TEACHERS AND WORKSHOP ASSISTANTS

If you are lucky you may have a counterpart teacher (newly trained teacher) or a workshop assistant to work with you. It will be your responsibility to train the teacher to teach Introtech in the integrated manner intended and to train the workshop assistant to organize the workroom for students' work.

The workshop assistants are usually from trade backgrounds and some of the more enthusiastic ones go on to make excellent, albeit unofficial, teachers. The balance that you establish between informal and formal methods of training your counterpart will be a very personal thing, dependent upon the personalities, confidence and experience of both yourself and your counterpart. The following are a few general guidelines:

- Share your ideas, enthusiasm or despair.
- Use your counterpart's local knowledge for ideas of projects, organization, local resources and teaching.

- Find out the methods of teaching and working they have formerly used.
- Don't try and set up things which will not be sustainable by your counterpart.

LEAVING A RESOURCE

When a teacher leaves a school it is their duty to leave a teaching resource package behind. This resource should be built up naturally over their time at the school so little extra work is needed to prepare it. The resource should include the following:

RESOURCE NOTES

Resource notes should contain all the information necessary for a particular lesson. The new teacher can use the same information in a way that suits them.

LESSON PLANS

Copies of lesson plans can be left but these are often written in a very individual manner and are not as transferable as resource notes. However, they do give a very good idea of the quantity and quality of work that can be covered in each lesson.

SCHEME OF WORK

A full scheme of work must be left with extra notes explaining how the scheme was taught, e.g. number of lessons, type of lesson, etc.

PROJECT WORK

Copies of all project drawings should be left with appropriate notes detailing methods of construction, cost of materials, time taken and effectiveness of project.

MARK BOOKS

The teacher must leave copies of all student marks as these will be needed by other teachers for continuous assessment. It is important they are written clearly and are understandable by all.

MATERIALS SOURCES AND LOCAL CONTACTS

During your time at the school you will build up a lot of contacts who can help you find tools, materials, etc. It is important that you list the names and addresses of these contacts and the addresses where certain materials are available.

VISUAL AIDS

All visual aids should be left. If possible, they should be numbered and the corresponding number written in the appropriate resource notes or lesson plan for which they are needed.

EXCHANGE WORKSHOPS

By talking to other teachers you will soon realize how many good ideas there are around. The problem is that teachers rarely get the opportunity to meet up with one another. Because of this, a vast amount of ideas are either being lost or the same ideas are being 're-invented' all over the place.

In order to try and alleviate this problem and make people's teaching more effective, exchange or skill workshops can be organized locally. The planning of these will be up to the interested parties involved; it may be half a dozen teachers who want to get together or the Ministry may be interested in making a contribution. Each workshop should have a theme, i.e., activity ideas, maintenance of tools, writing schemes of work, etc. Teachers who are experienced in a particular area may act as tutor during the workshop.

The possible effects these workshops can have on the teaching of introductory technology are endless and should be encouraged at all levels. It is through workshops such as these that efficient methods of teaching can be passed on to teachers all over the country.

10. FUNDING, EXTRA WORKROOM USES AND SCHOOL VISITS

FUNDING

Funding is possibly the biggest problem faced by teachers of introductory technology. When trying to get funding for the department you may get frustrated by what seems like lack of interest on the part of the people who should be supporting you. *Don't give up.* It is important that you try as hard as you can to get money from the Ministry and Principal. The aim should be to make the ministries aware of the financial commitment they have to make for introductory technology to become successfully established within the country.

The other methods of raising funds mentioned here should really be used only for extra funding and not direct funding, which is the responsibility of the ministries.

NB: all methods of funding used *must* have the Principal's approval.

WHERE TO LOOK FOR FUNDING

MINISTRY

The Ministry of Education controls the majority of money allocated to the development of introductory technology. It is important that the State Ministry is aware of the needs of your school. Don't wait for the Ministry to approach you . . . tell them what the Introtech department needs.

Once you have decided what the department needs, you should make a formal application for funds through the Principal, then through the local zonal office and then to the Ministry. It is important to find out the school reference number for dealing with the Ministry as it may make things quicker. Getting funds from the Ministry can be difficult, but by visiting the Ministry regularly, explaining the benefits the money will give and showing enthusiasm, funds should be allocated (even if it's just because you have forced them into submission through continual pestering).

PRINCIPAL

If the Principal has been given money from the Ministry, you must

approach the Principal for funds, remembering that most other departments in the school will also be applying for funds.
Always make a formal application in writing listing everything the department needs and the cost. Once you have made the application, keep asking the Principal about the money until the day you actually get it.

PARENT–TEACHER ASSOCIATION (PTA)

In schools where the PTA is active this is a very good source of funds. When asking the PTA for funding you should:

- Call a meeting of the PTA – or find out when the next one is.
- Make a formal request for funds, explaining exactly why the department needs funding.
- Where possible, hand out copies of your written request (it looks professional and will not be forgotten so easily).
- Make sure a copy of your request goes into the PTA file and your own file.

COMMUNITY

In rural community schools you can call a meeting of the community. You should write letters of invitation to the Chief and village elders asking for the intended meeting to be announced in the village, either at council meetings or by the village crier. The format of the meeting should be the same as for the PTA.

LOCAL INDUSTRY

If there are small industries in your area it may be worth approaching them for assistance; some teachers have been given considerable amounts of materials by them. It is always very important that you make contact with local industries as these are the places many of your students may work in, in the future.

OLD STUDENTS' ASSOCIATION

Old students will often give money to a project that will improve the school. You should find out from the Principal if your school has such an Association.

LOANS OF EQUIPMENT

Although this is not a very common occurrence, some teachers have been given equipment on loan from technical colleges that have excess tools. However, if this does happen it is essential that all the people involved realize it is *only* a loan and funds still have to be raised through the official channels.

ROTARY/LIONS CLUBS

Many towns have Rotary/Lions clubs that may be willing to give support to your school.

SELF-FUNDING BY STUDENTS

The majority of students will be able to pay a small amount for the project they are making. By charging the students, the department can become almost self-sufficient as far as materials are concerned. However, other funds will always be needed for items such as hacksaw blades, sandpaper, etc.
If you decide to charge students for project work it is important to remember the following:

- Students must want to make the project.
- Always collect money before the students start work.
- Record all money collected.
- Show students what has been bought (or take some students with you to buy materials).

INTRODUCTORY TECHNOLOGY CLUBS

An Introtech club may be set up to encourage students to learn about technology outside of school hours. In some areas local and national competitions are held between schools.

SCHOOL FUNCTIONS AND LEVIES

In some schools where there is no PTA the students are levied when money is needed for certain projects around the school. If this is the case in your school you should consult the Principal about levying the students. If the school has Open Days, various activities can be arranged that will help to raise money for the department: charity football matches, dancing, exhibitions and competitions between nearby schools.

LAUNCHING

Often when a new department is about to open in a school a launching is held. The aim of the launching is to inform the community of the development within the school and to raise money through donations for the running of the department. Considerable amounts are often raised at such events. If the department is new you should discuss the possibility of a launching with the Principal.

ADULT CLASSES

Adult classes should only be considered if the teacher is sure they will have enough time. If time is available they are a good way to improve the general public's opinion of technical subjects and to raise a small amount of money for the department.
The following is a checklist for setting up an adult class:

Fig. 10.1 Launching

(a) Decide what subject(s) can be offered.
(b) Contact the relevant people: Principal, Chief, tradesmen, villagers, as to the support you are likely to get and the interests and desires of your potential students.
(c) Get authorization from your Principal for a specific class.
(d) Inform through the Chief, churches, school and bars or meeting places what type of class is available, who is eligible, and when and where enrolment will take place (with emphasis placed on 'first come, first served').
(e) At enrolment, register names and addresses as people arrive (this is very important to avoid any arguments later).
(f) Discuss in detail what the class is to do and get consensus opinion of what is wanted.
(g) Rules must be clearly stated and acceptance of them agreed.
(h) Strike out those no longer interested or eligible. Give a start date and time.
(i) As a result of the meeting you should now know in detail what tools, materials, visual aids, lesson plans, etc. are required.

COMMUNITY USE

If the school is a community school you may be approached by members of the community asking if some of the local craftsmen can use the equipment. If you decide to allow people to use the workroom and equipment you must keep the situation under tight control. Always remember the following:

(i) Students come first at all times.
(ii) The workroom is not a commercial venture.
(iii) No tools should leave the school compound.
(iv) If a generator is being used, the operator should pay for the upkeep (oil, diesel, servicing, etc.).
(v) Agree on forms of payment from the beginning (materials etc.).
(vi) The idea must be officially approved by the Principal and community.
(vii) Everyone involved must understand the situation fully.

The following are some ideas of how the workshop can be used to help the local craftsmen.

HAND TOOLS

Many tool-kits have a full range of tools covering various crafts. These tools could be used by local craftsmen, enabling them to do various jobs that they could not do previously.

MACHINE TOOLS

Craftsmen can be trained to use the machine tools (if they do not already know), and local people already trained can rent machine time to produce saleable items. (Metal- and wood-lathes are particularly useful in this respect.)
The circular saw and planer would help carpenters tremendously in the preparation of timber, and the tyre remover would be invaluable to local tyre-repairers. (The teacher should be vigilant as to how the machinery is used as there is no insurance against broken cutters.)

BATTERY CHARGER

Some tool kits are supplied with a battery charger. If the school has electricity this should never be idle – by contacting local mechanics you will soon have a steady flow of batteries to charge and a steady income.

WELDER

If there is no electricity in the village but the school has a generator, then a local welder could come and work from the school.

EFFECTS OF LOCAL CRAFTSMEN ON STUDENTS

If craftsmen come to work at the school you may be able to incorporate their activities into your scheme. The following are some ideas on how the students can be involved:

- Students can watch the craftsmen working.
- Projects can be organized around the work they do.
- Some students may be able to work with the craftsmen.

HOSPITAL AND DISABLED WORK

You may be approached by a local hospital to help make some aids for disabled patients. If you have time, this is always a very good thing to do and it may be worth visiting the nearby hospitals, especially the leprosy hospitals.

IMPROVING SCHOOL ENVIRONMENT

There will always be parts of the school that need improvement: water tanks in need of repair, latrine covers to be made, etc. You may be asked to help repair items in the school.

Although it is not the teacher's job to be the repair-man, if you have time or the work fits into the students' scheme, obviously it will benefit the school, e.g. one teacher had students making latrine covers for the school. This was an ideal project as it fitted perfectly into the syllabus and helped the school environment considerably.

SCHOOL VISITS

The teacher should try and plan at least one visit for each class while they are learning introductory technology. If possible, the particular place visited should be appropriate to the part of the scheme being covered at the time. Here are a list of places that will be of interest to students:

- Local craftsmen: students can learn a tremendous amount from watching these people at work. (Make sure they receive lots of good publicity.)
- Industries/factories: although not found in every town, where a school is close by they are worth a visit.
- Technical schools: VTCs etc. – it is very important that the students who want to follow a technical subject visit these schools before making a decision on their career.

NB: when organizing a visit, always go through the Principal and put everything in writing.

11. MATERIALS

Obtaining materials for your department can often be a problem, mainly due to lack of funds. However, a tremendous amount of material can be found locally. Tin cans, plastic bottles, clay, bamboo and paper (if you make your own) are always available for little or no cost. Lots of other materials, especially wood, metal and wire scraps are fairly easily obtained. The teacher will constantly have to reassess his or her preconceived ideas as to which materials can be used for a particular job. For example, what can a teacher use if he or she has no blackboard or blackboard paint? Maybe an old door, table top, part of an old water tank or smooth one of the walls in the classroom. As for paint, charcoal can be mixed with gum arabic or watered-down glue. The teacher must learn to be inventive. If your desired material is not available, don't give up – think of an alternative; the alternative idea may turn out to be better than the original! This chapter explains how materials can be found and collected. It also gives advice on buying items from hardware stores and finishes with a section on appropriate material ideas.

COLLECTING MATERIALS

The people who know most about what materials are available and where to find them are your students, colleagues at school and local craftsmen. By visiting mechanics, welders, potters, weavers, tinkers, etc. you will soon know where materials can be bought and found. Don't forget the school compound and storeroom; there are often many broken desks and stools that can either be repaired or used as material for projects.

COLLECTING BOXES

A very good method of collecting materials is to have material boxes in the workroom (Fig. 11.1). These can be labelled – metal, wood, plastic, etc. Posters can be put around the school asking for materials to be brought to the workroom.

NOTE OF WARNING

Try and find out where students are collecting their materials. One enthusiastic student brought his teacher lots of round tube for which the teacher was very grateful. It was only when the

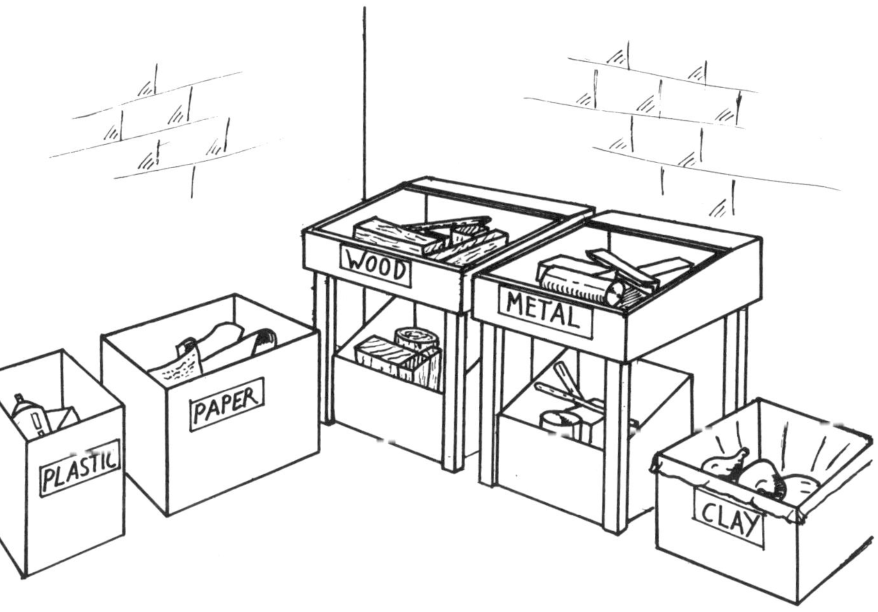

Fig. 11.1 Collecting boxes

electricians came that the teacher realized all the conduit had gone from several classrooms!

HARDWARE/MATERIAL STORES

The majority of small towns will have some type of hardware/ material store. These often turn out to be an Aladdin's cave, selling things you never imagined would be available.

When buying tools and equipment you must be very careful that what you are buying is original, as many items have at least two copies. The following are a few of the items that are copied and how you can find the original:

HACKSAW BLADES

There are many different hacksaw blades available. Beware of any that are very cheap and be very careful if buying Ellipse as there are many copies. The only one you can always guarantee is the 10″ Ellipse. Basically the more expensive the blade, the more chance there is of it being original.

NB: it may be worth buying some cheap ones and seeing if it is possible to re-harden and temper them.

BRAZING RODS

Buy the most yellow-looking ones. Many seem to have a very high melting point with the students work melting before the rod!

SOLDER

It should be fairly shiny and not too heavy. Sometimes pure lead is sold.

GLUE/PVA

Usually the most expensive brand is the best. *Always* open up the pot and smell the glue – it should smell acidic. It it smells like coconut milk do not buy it – it probably is!

PAINT

Always ask to look at the contents and check it's not too watery. (See notes on making paint.)

ABRASIVE PAPER

Different brands vary enormously. Don't bulk-buy one brand, but experiment with different samples first. (See notes on making sandpaper.)

These are just a few examples. The important thing to remember is to check what you are buying before you pay. If you are unsure as to what you should be checking, ask the local craftsman what the original should be like.

LIST OF COMMONLY AVAILABLE MATERIALS

NAME OF ITEM	TYPE OF MATERIAL	NOTES
Leaf springs	High-carbon steel	Tool-making. When hardening quench in oil as it will crack in water
Coil springs	High-carbon steel	Tool-making
Car panels (door skins)	Mild steel (maybe aluminium)	General sheet-metalwork
Brake shoes	Aluminium	Good for melting down for casting
Engine mountings	Rubber & metal	Large mountings make good anvils
Car bumpers	Mild steel	General work and good formers for sheet-metal

NAME OF ITEM	TYPE OF MATERIAL	NOTES
Pistons	Aluminium	Casting
Petrol tanks	Mild steel	Sheet-metalwork
Oil drums	Mild steel	Very good for the basis of a forge
Water tanks	Mild steel	Old ones provide a good source of material
Desks and chairs	Mild steel and wood	The tube and wood can be used for a variety of projects
Tin cans	Thin sheet	One of the best material sources available
Car seats	Wire from springs, foam and cloth	Always useful to have around
Coat-hangers	Aluminium	Aluminium coat-hangers make excellent rivets
Bicycle and machine spokes	Steel	These can be made into very simple drill bits, very good for wood
Packing crates	Pine, metal	A good source of wood. The metal strip can be used to make knives
Inner tubes	Rubber	Strips can be used for making toys powered by elastic and tool handles
Plastic bottle containers	Thermo-plastic	Plastic can be melted down and formed into any shape needed
Calabash		A base for many projects, lamps, containers, etc.

NAME OF ITEM	TYPE OF MATERIAL	NOTES
Bamboo		Can be incorporated into many projects: torch body, puppet, etc.
Clay		Found everywhere and a very versatile material. (See extra notes.)
Gum arabic	Resin from the acacia tree	Very good base for making own paints and glues
Charcoal	Charcoal	Apart from obvious use for forge, it can be used as a colouring for paint mixed with gum arabic
Paper	Paper	Papermaking is a very good project. (See later notes.)
Sand	Sand	Can be used for making very good sandpaper. Needed for casting
Glass bottles	Glass	Always useful to have around for various projects
Animal hair	Hair	Hair from goats, dogs and cow tails etc., can be used to make very good paint-brushes
Animal bones	Glue and Charcoal	Old bones from the market can be bought and boiled up to make glue. (See later notes.) Animal bones also produce very good carbon-rich charcoal for case hardening
Car tyres	Rubber	Can be used for making sandals

NAME OF ITEM	TYPE OF MATERIAL	NOTES
Radio speakers	Magnets	Visit your local radio repairer
Starter motors	Copper wire	Good for electro-magnets
Bedsprings	Wire	Can be used as welding rods for gas welding
Magnetic strip from a fridge door	Magnets	The sealing strip on the inside of a fridge door is magnetic

COMMONLY AVAILABLE MATERIALS AND GENERAL MATERIAL IDEAS

Most towns will have at least one person selling timber, metal and cement. The materials listed here are the most commonly available.

All the ideas shown here have been tried and tested; however, it is important that you experiment – you may find a better method! If you don't quite understand a process, get stuck in and have a go – it will soon become clear.

METALS

Round bar The majority of round bar is concrete-reinforcing rod (locally known as Protector in Nigeria). It is available in 30ft lengths in the following diameters: 6mm, 8mm, 10mm, 12.5mm, 16mm, 20mm.

NB: 12.5mm, 16mm and 20mm harden fairly well.

Flat bar Black mild steel available in 20ft and 40ft lengths of the following sizes: 12.5 × 3mm, 25 × 8mm, 50 × 6mm.

Mild steel sheet Available in 16g, 18g, 20g, all sheets are 8ft by 4ft.

Galvanized sheet Available in 16g, 18g, 20g, all sheets are 8ft by 4ft – very expensive.

Zinc roofing sheet This zinc-plated roofing sheet is commonly available and can be flattened for projects.

Angle iron Available in 20ft lengths of the following sizes: 20mm, 25mm, 36mm, 50mm, 75mm and 100mm.

Round and square tube Available in 20ft lengths of the following sizes: round – 16mm, 25mm, 36mm; square tube – 25mm, 16mm, 25mm, 36mm.

Wood Obeche (physically soft), iroko, mahogany. There are many different woods available; the three listed are the most common. Sizes are usually 12in × 1in × 16ft and 2in × 2in × 16ft although other sizes can be found.
Plywood – available in 8ft and 4ft sheets of the following thicknesses: 6mm, 10mm, 12.5mm, 20mm.
Veneer – various veneers are available.

METAL IDEAS

Finishing Much of the metal used will be steel and, if left bare, it will soon go rusty. Painting is an obvious answer but if paint is not readily available, oil-blacking provides a pleasant and hard-wearing finish.
To oil-blacken a piece of metal, heat it until it is a very dull red colour when looked at in a darkened corner, then submerge it in old engine oil, remove it and allow it to cool.

HEAT TREATMENT

Hardening Only medium- and high-carbon steel can be hardened by the following method (all car springs are high-carbon steel). Heat the part of the tool that needs hardening until it is bright red (not orange) and quench in water or oil.
NB: if hardening high-carbon steel (leaf springs), always quench in oil as it may cool too quickly in water and crack. If the item being hardened is a chisel or plane blade, put it into the oil vertically otherwise it may distort.

Safety Take extreme care when using oil as sometimes it may ignite; to avoid this, make sure the whole tool is submerged.

Tempering This is done to remove some of the brittleness which is a by-product of hardening. Clean the cutting edge of the tool until it is shiny; heat it gently until you see colours appear on the metal (you should do this in a shaded area so the colours are easy to see); each colour represents a temperature; to temper a chisel or plane-blade the colour needed is purple to dark purple. Once you have seen this colour appear on the tool's cutting edge, quench it in water or oil.

Case hardening Low-carbon steel cannot be hardened in the usual way and extra carbon must be added before it can be hardened. The standard way of adding carbon to a piece of low-carbon steel is to place the item to be hardened in a steel box and surround it with carbon-rich charcoal (animal bone produces the richest carbon). The box is then placed in the furnace and heated at about 800° for a few hours; during this time the metal absorbs the carbon. If no box is available the item to be hardened can be surrounded by charcoal and wrapped in clay; this is then put into the furnace.

If only a small area needs hardening, such as the tip of a screwdriver or chisel, then the following method may be used:

Heat the metal until yellow and rub quickly on a piece of wood; repeat this a few times. When the hot metal is rubbed on the wood, the wood burns, which produces carbon, and the metal then absorbs the carbon.

NB: much of the metal used will be of an unknown quality and it is advisable to experiment with hardening, tempering and quenching liquids on scrap metal before heat-treating your finished work (it's very depressing when your plane blade cracks after ten hours of work!).

MAKING DRILL BITS

Simple drill bits useful for wood can be made out of bicycle or motor bike spokes by flattening the end of the spoke until it is a diamond shape, as shown (Fig. 11.2).

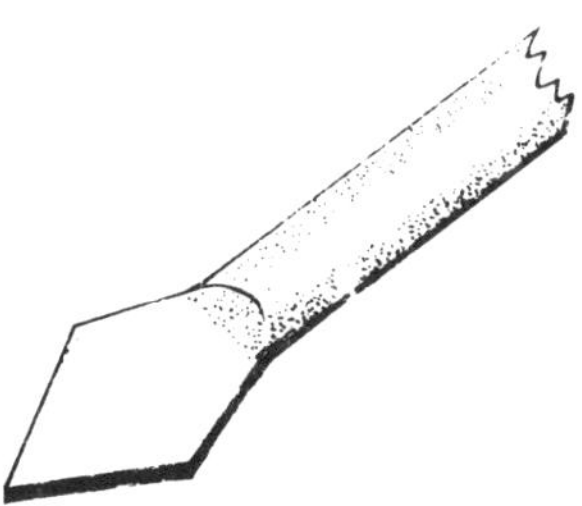

Fig. 11.2 Making drill bits

CLAY

Clay is an excellent material for students to work with; not only is it free and widely available but it can be incorporated into many different topics, from building to jewellery-making. Clay is found all over the country in river beds and banks, in ponds, marshy places and termite mounds. Children always seem to know where to find it, or ask potters – nearly everyone has some idea. Take time to look at all local clay sites as there will be less preparation to do on clean, stoneless clay.

COLLECTING AND PREPARING CLAY

It may be possible to find stone-free 'plastic' clay which can be used as dug; if this is not possible, collect the clay in a dry state and pound to powder, take out the stones, then mix with water until of modelling consistency. For making things, clay needs to be rollable into a sausage shape and bent into a circle without much cracking; if the clay is slimy or too sticky, add fine sand or, for larger work such as bricks, add chaff or dried dung. If the sausage cracks when bent, add the powdered clay to a bucket of water until it is the consistency of thin cream, pass through a sieve to screen out the largest particles of sand, allow to settle and siphon off water; then allow to dry until of modelling consistency. Clay preparation can be time-consuming but if carefully planned it becomes an instructive and fruitful practical lesson providing free material for many projects.

STORING CLAY

Dry as dug: this should be kept dry and left in a heap near the workroom.
Powdered: store in boxes, bins or sack.
Plastic as dug: if it is stony, dry out completely. If it is stone-free and ready to use, store as below.
Small pieces of clay should be wrapped tightly in polythene. Larger pieces of clay should be stored in some type of container such as a bin or pit. Cover in wet sacking and polythene, and cover with a tightly fitting lid. A pit or bin sunk into the ground will keep very cool.
Clays vary enormously both in colour and texture but all clays must be heated to a minimum of 500–600° before they are changed to pottery.

MAKING PAINTS

Charcoal, coloured clays, earths and chalks can be used for cheap, easy paints, inks and watercolours. The material used should be crushed and sieved until very fine and mixed with gum arabic or other glue and water. The ink can be dried in small trays and will become water-coloured tablets.

GLUES

Simple paper glues can be made from cassava paste, cornflour (pap) and any type of flour/starch and water.
A stronger glue can be made by pouring a small amount of petrol on polystyrene. The resulting glue is a contact adhesive and is fairly good for joining wood.

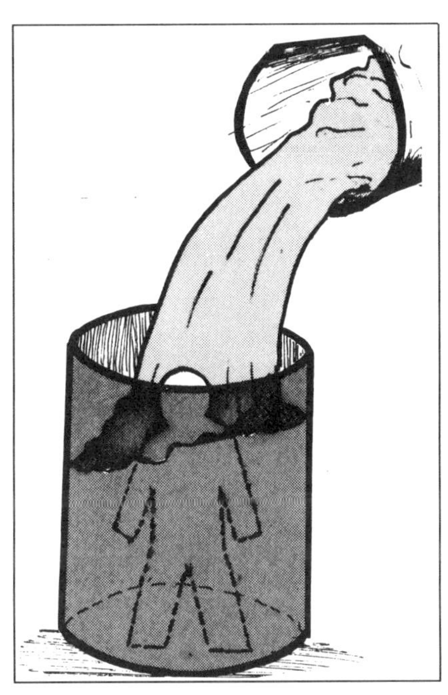

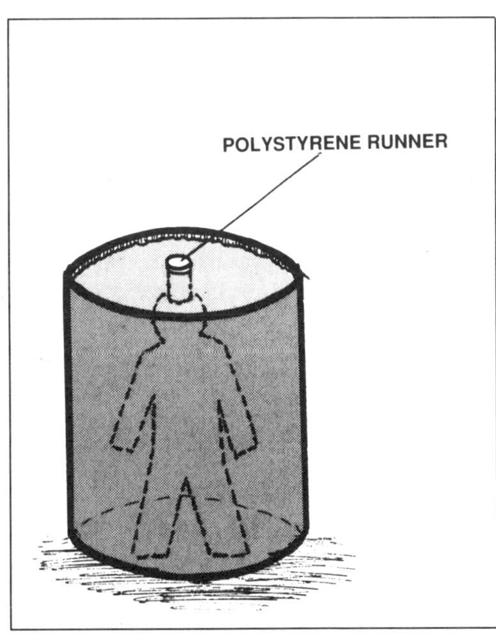

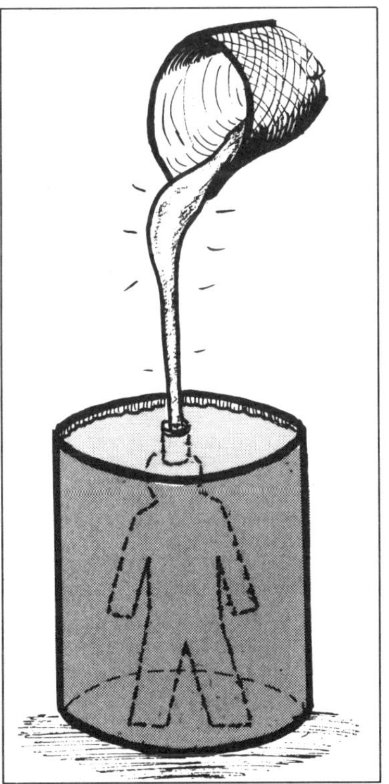

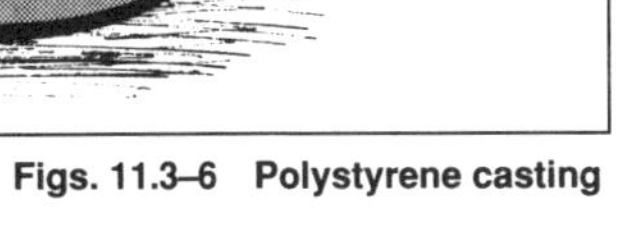

Figs. 11.3–6 Polystyrene casting

Animal bones and skin can be boiled up to make glue, but they have to be boiled for a very long time.

MAKING SANDPAPER

Very good sandpaper can be made by simply sieving sand and glueing it on to paper. Cement bags provide a very good source of paper. It is very important that the glue used is a good-quality PVA. By making different-sized sieves, fine, medium and rough papers can be made.

SAND FOR CASTING

Most sand found will be good for casting. Termite sand is especially good and even sandy soil can be used. The important thing to remember is that it should be possible to mould the sand into a ball without it cracking but at the same time not be too wet.

RELEASING AGENT

Cement dust or fine ash can be used as a releasing agent for sand casting.

POLYSTYRENE CASTING

Not everyone will have access to pieces of polystyrene but if you do come across some, polystyrene casting is wonderfully simple. Below is a basic outline of how it's done. NB: do this in a well ventilated area as the fumes can be dangerous.

1. Carve your design out of polystyrene.
2. Place the polystyrene inside an empty tin, making sure there is at least 50mm clearance between the polystyrene and the edge of the tin (Fig. 11.3).
3. Pour in fine, *dry* sand until it just covers the top of the design. Gently tap all around the tin with a piece of wood; this will allow the sand to fill every part of the mould (Fig. 11.4).
4. On the top of the design you should put a runner (this is where the aluminium will be poured); this can just be a straight piece of polystyrene. Pour in more sand until just the top of the runner can be seen (Fig. 11.5).
5. Pour in the molten aluminium (Fig. 11.6).
6. Allow to cool and remove the casting.

What happens is that the molten aluminium dissolves the polystyrene, giving off certain gases which keep the dry sand in place.

JELLY DUPLICATOR

The jelly duplicator is an extremely effective method of printing, producing up to 80 copies from one stencil. There is a commercially available kit which is excellent, allowing the teacher to make 8,000 copies before the printing compounds wear out (see Resource section). If the teacher is unable to obtain one of the kits it is possible to make your own.

INSTRUCTIONS

You can use either gelatine or jelly (banana and orange flavours are good). If using jelly, only use a small amount of water so that it sets firm; if using gelatine (which is best), follow these instructions

2 sachets (22g) gelatine:
2 cups (½ pint) water:
4 tsps sugar;
100 mls glycerin/glycerol.

1. Boil water and place in pan.
2. Add gelatine and stir vigorously.
3. Add sugar and stir until dissolved.
4. Add glycerin and stir.
5. Pour into A4-size tray.
6. Remove any bubbles before allowing to set.
7. Make a stencil using banda carbon, water-soluble pen or ink. (Banda carbon is the most efficient but may be hard to obtain in many countries.)
8. Lay the stencil on the jelly for 40 seconds, then peel off.
9. Lay your paper on the jelly (turn up corner for easy removal); peel off immediately.

NOTES

The type of pen used is very important; it must have a high concentration of ink; many felt pens are too weak (purple is the best colour). Special pens are available (see Resource section). One of the best alternatives to using pens is to use gentian violet (available from chemists all over the world). Twenty copies can easily be produced – it is especially good for drawings and charts.

If it is found that the paper is sticking to the jelly, it means the jelly is too dry and needs dampening; to do this, just wipe across the jelly with a damp cloth.

The jelly can be re-used by either melting down or letting the ink sink to the bottom. By letting the ink sink to the bottom and keeping the surface dry and covered, the same jelly can be used over and over again. Alternatively, if using 'real jelly', buy some fruit and eat it!

PAPER MAKING

Making paper is a very effective project covering many different subject areas. It's also a valuable resource for the teacher with a

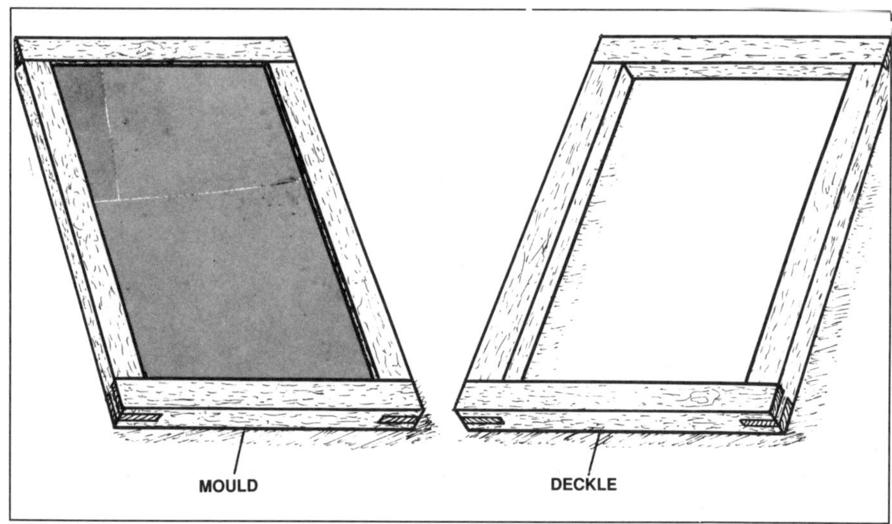

Fig. 11.7 Papermaking: items needed

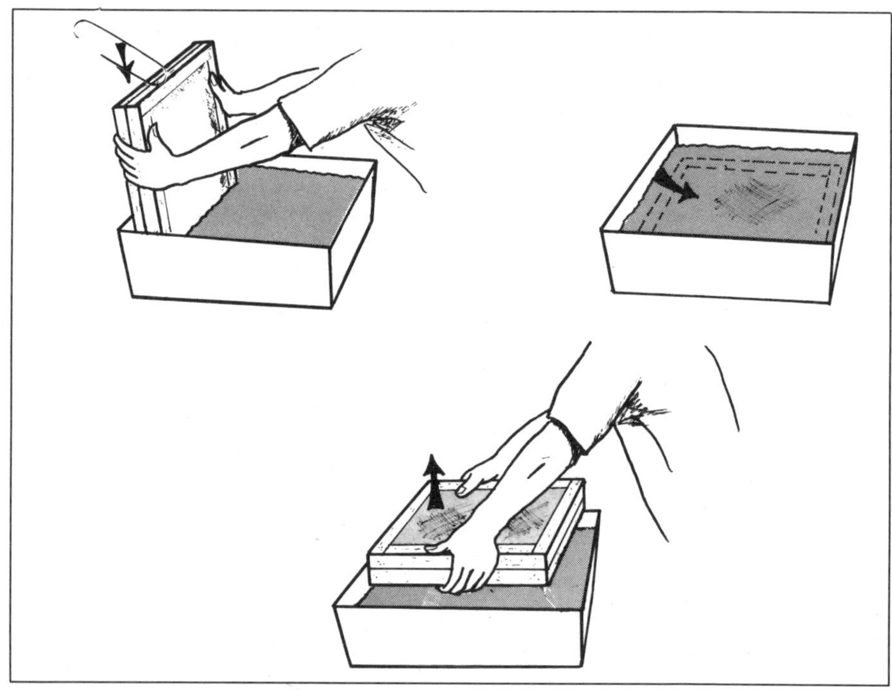

Fig. 11.8 Making the pulp

Fig. 11.9 Making the paper

paper shortage. The quality of paper produced depends greatly on the type of materials used.

ITEMS NEEDED

1. Two wooden frames (Fig. 11.7); the inside of the frame should be the size you want your paper (A4 = 30 × 21cm); the wood used should be about 1 × ½″ (25 × 12.5mm). Across one frame, stretch as tightly as possible a mesh such as mosquito netting, cloth sacking or any loosely woven fabric (not cotton); this frame is called the mould while the other frame is called the deckle.
2. Cloth for pressing: non-textured cloth is needed for putting between the sheets of paper during pressing. Shirt-collar stiffener is good, as is felt if it can be obtained.
3. Press: two pieces of plywood and clamps or heavy weights make an efficient press.
4. Water container: this must be large enough to fit the mould and deckle in.
5. Pounder: for pounding the pulp.

RAW MATERIALS

Any kind of waste paper may be used – newspapers, old test

papers, etc. If raw materials such as banana leaves, grass or corn stalks are used, the fibres must be broken down by soaking or boiling in caustic soda and water (soap-makers use caustic soda – many towns will have soap-makers); once the pulp has been made, the process is the same as for old paper.

MAKING THE PULP

1. If using old paper, tear into small pieces and soak in water for several hours.
2. Take a handful of the soaked paper and pound it until there are no recognizable pieces of material left. Repeat this process with the rest of the soaked material.
3. Once all the material has been made into pulp, put half of it into your water container and add about four parts water (you will need to experiment with quantities of water). The remaining pulp is used for topping up the mixture as each sheet of paper made reduces the strength of the pulp.

MAKING THE PAPER

1. Place the deckle on top of the mesh-covered side of the mould and hold them firmly together.
2. With the deckle uppermost, dip into the mixture (Fig. 11.8), starting at the far side and bringing forward until the mould and deckle are horizontal and just covered by the mixture. Hold steady for a moment and lift out the whole deckle from the pulp swiftly.
3. Remove the deckle.
4. Put two pieces of pressing cloth on to one of the press boards.
5. Turn over the mould and press the pulp on to the cloth (Fig. 11.9).
6. Hold the mould and board tightly together and tilt them so that excess water can run off.
7. Lift the mould off.
8. Continue this process, stacking the paper between layers of cloth. Keep topping up the mixture with the remaining pulp.
9. When the last piece of paper has been made, cover it with cloth, put the other press board on top, then either clamp the boards together or put a heavy weight on top, Let the water drain out for about 30 minutes.
10. Remove the weight and peel off each layer of paper still on its cloth and leave it to dry for a few hours.
11. Remove the cloth and allow the sheets to dry for a few hours.
12. Stack the papers between the two boards under a light weight and leave for a day or two until perfectly dry.

NB, you will have to experiment with drying times as they depend upon the climate.

PAPIER MÂCHÉ

Use flour/starch and water paste. If you are using something as a mould, i.e. a calabash, make sure you put Vaseline (or similar) on the mould first so that the item being made can be released easily.

MAKING SWITCHES AND BULB HOLDERS

1. BULB HOLDERS (Fig. 11.10)

Easily made by straightening out a paperclip, and twisting it around a pencil or something else the same size as the bulb thread. The bulb can then be screwed in. The positive wire should be attached to the paperclip, and a small plate or piece of tinfoil should be attached to the negative wire, which should be positioned under the bulb so that the bottom of the bulb touches the plate.

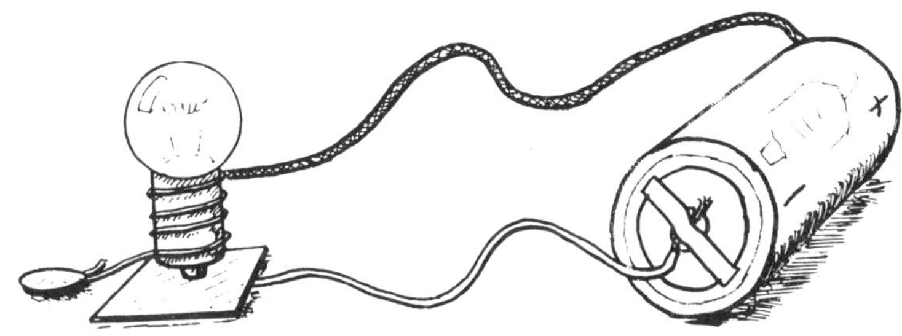

Fig. 11.10 Bulb-holders

2. SWITCHES (Fig. 11.11)

Attach wires to two small nails, screws, etc. Use another piece of wire or paperclip to cross the two points, thus making a circuit.

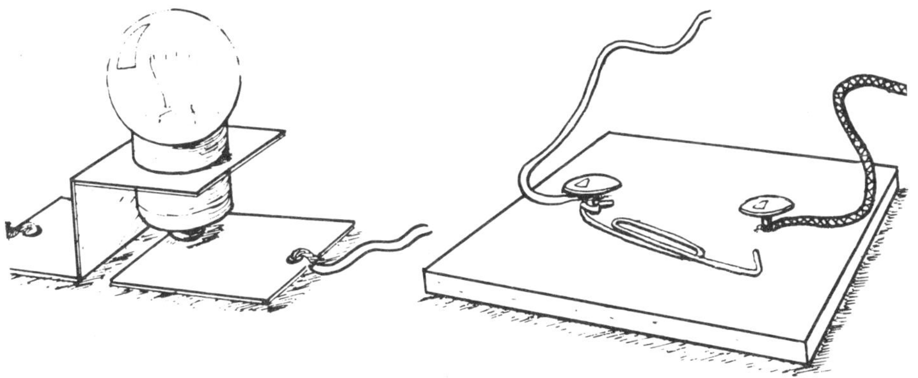

Fig. 11.11 Switch

BRICKMAKING

Brickmaking is both a very useful skill to learn and an excellent integrated project, incorporating woodwork, metalwork, building, drawing, energy (firing), etc. First a wooden brick mould must be made; at its simplest this is just a wooden box with no top or bottom. Your material is then put into the box and compressed as much as possible; chaff or dung can be added to the brick clay to avoid cracking when firing.

If making many bricks, stack them evenly in a circle leaving a half-metre hole in the centre (Fig. 11.12). Fill the centre hole with dried leaves and grasses. Light a small fire for an hour or so to warm the bricks thoroughly, then cover them completely with grasses, sticks and logs and allow to burn overnight.

NB, if a large amount of bricks are to be made it may be worth making a brick-making machine (one type is called the Cinva ram in Nigeria). Ask local builders how.

Fig 11.12 Brickmaking

12. WORKROOM ACTIVITIES AND PROJECTS

This chapter deals with activities and projects that the students can carry out during their lessons. The most important point to remember is that Introtech should be 'hands on' – the student doesn't always have to be making something; doing something is just as valid.

ACTIVITY LESSONS

The teacher should try to make each lesson an activity lesson – a finished product is not always necessary as long as the student is learning and understanding from the activity. When setting a project, try to ensure that it is the most effective method available for getting the subject across to the student.

SETTING PROJECTS AND ACTIVITIES – SET EXERCISE

The teacher sets a specific project, e.g. 'make a bottle opener like this', and has control over the situation, understanding all the skills involved and possible pitfalls. Worksheets can be set that will allow the students to work at their own pace and learn by discovery. Setting a project in this way gives the students confidence as well as transferring skills which they can continually build on.

SETTING PROJECTS AND ACTIVITIES – PROBLEM SOLVING

This method allows for individual interpretation. For example, students may be given a pile of assorted materials and asked to make a device for opening bottles. It encourages creativity and imagination amongst students, allowing them to use many different materials, tools, and concepts, thus covering wide areas of the syllabus. Co-operation is encouraged while discovery and questioning become important. It should be noted that students need to learn how to work by solving problems. It may be found that this method of teaching will improve after gradual introduction. Problem-solving activities require a lot of organization if an open brief is given, and it is often advisable to put some constraints on the project.

POINTS TO REMEMBER WHEN CHOOSING A PROJECT OR ACTIVITY

1. The standard and pre-knowledge of the students must be taken into consideration. The project may take more time than you think.
2. Each successive project should build on the students' past experiences in a logical sequence.
3. Projects should be chosen that allow for a multi-disciplined approach.
4. The project or activity must be relevant to the scheme being worked to.
5. The project or activity must have been carried out by the teacher before it is given to the students.
6. The project or activity should be interesting and fun in order to motivate the student.
7. If educational toys are being made, students should be encouraged to take them home to introduce to their younger brothers and sisters.
8. The cost of the project must be within the means of the students.

THINGS TO MAKE AND DO

Below are a few ideas of projects and activities that students can carry out. Each project or activity has a list of subjects covered; this is to assist the teacher when writing the scheme of work. Do not take the drawings too literally; if you do not have the materials mentioned, think of an alternative. In many of the following projects artwork, in the form of patternwork and design, will be required. Try to encourage originality; help students by displaying their own traditional designs, and patterns from their own cultural background.

NAME-PLATES

Subjects covered Measurement, shapes, materials, tools, design, finishing techniques.

Notes With metal, names may be written by using a centre punch. With wood, names could be burnt in, and with plastic, either painted, carved, burnt or moulded.

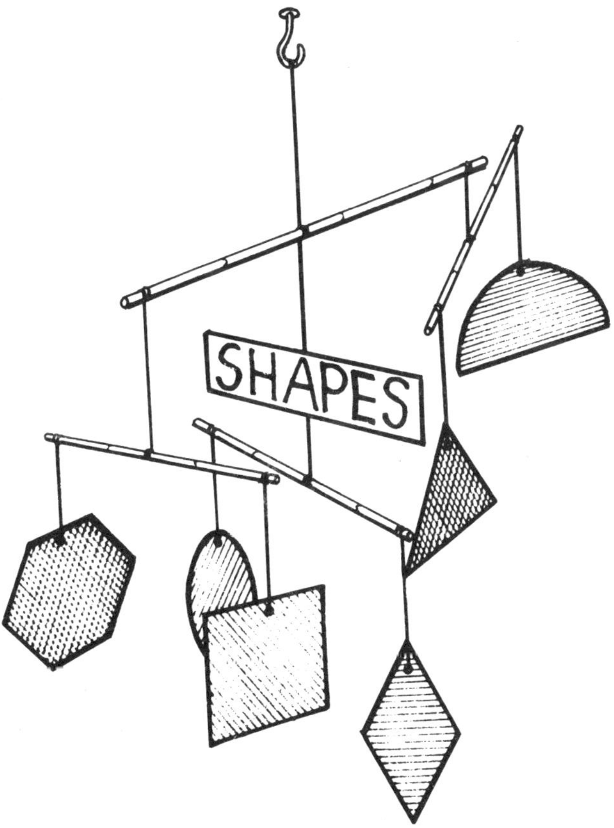

Fig. 12.1 Mobiles

MOBILES Fig. 12.1

Subjects covered Measurement, shapes, materials, tools, technical drawing (TD).

Note For first-year students this is an excellent project for teaching the basic skills mentioned above.

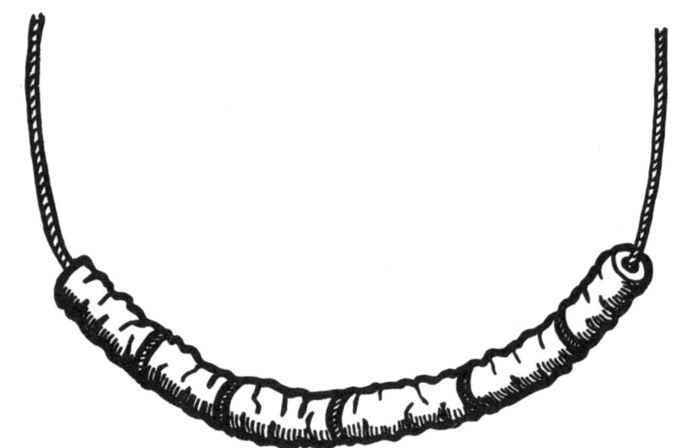

Fig. 12.2 Clay beads – jewellery

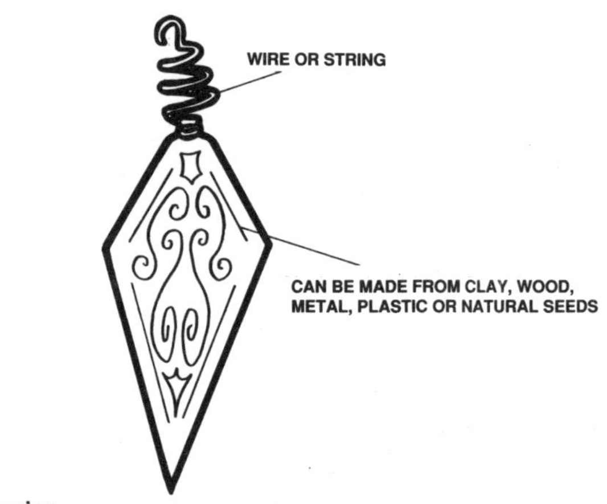

Fig. 12.3 Ear-ring

JEWELLERY/BADGES Figs 12.2, 12.3

Subjects covered Shape, measurement, sketching, design tools, marking out, finishing processes, joining techniques.

Notes Very good beads can be made from clay; roll the clay into small balls or any shape the student wants and make a hole through the clay with a piece of wire. Ear-rings and bracelets are easily made from wire and tin.

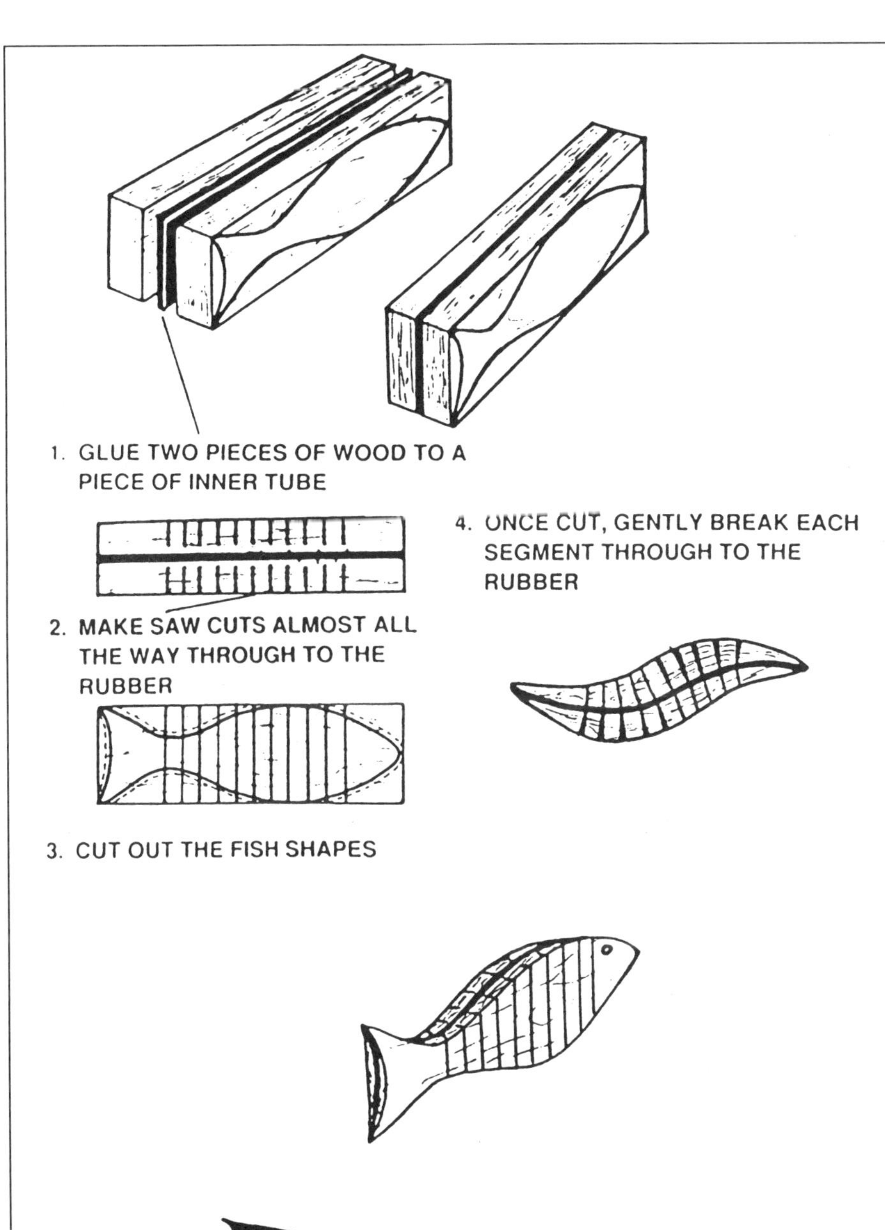

Fig. 12.4 Wiggly fish

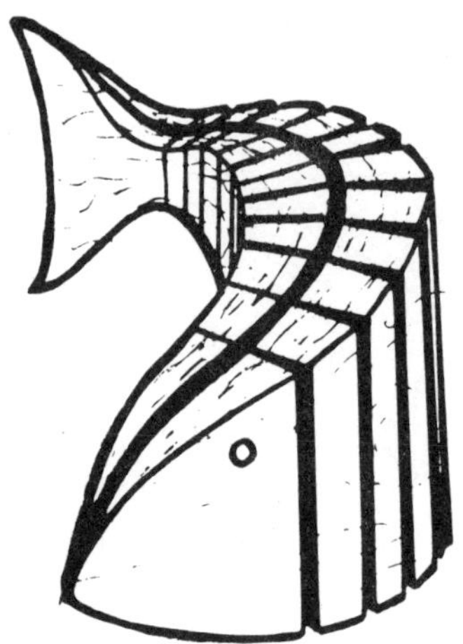

Fig. 12.5 **Wiggly fish/snake**

WIGGLY FISH/SNAKE Figs 12.4, 12.5

Subjects covered Materials, measurement, tools, sketching, design, glues.

Fig. 12.6 **Balancing toy**

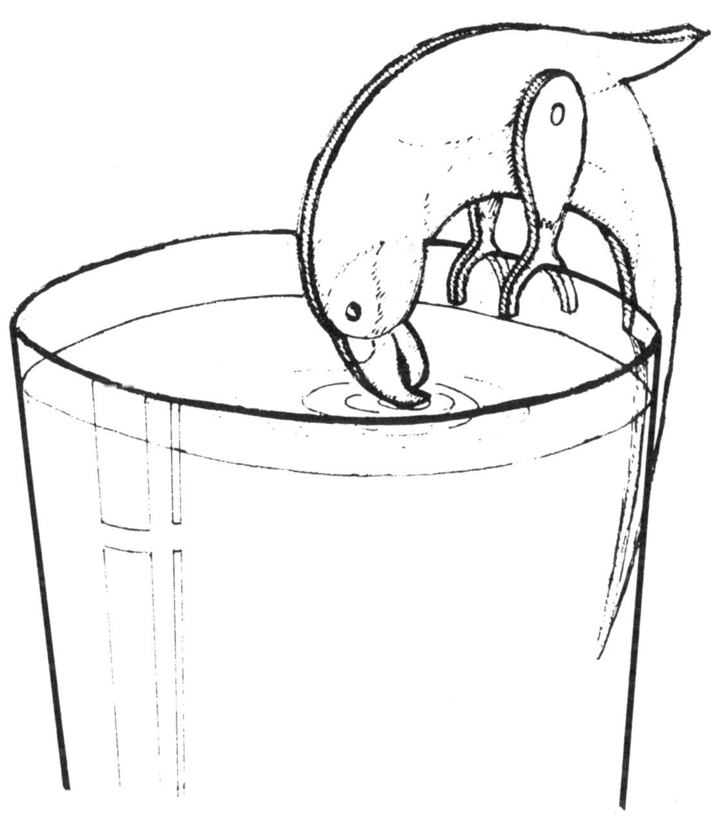

Fig. 12.7 Balancing toy (parrot)

BALANCING TOYS Figs. 12.6, 12.7

Subjects covered Sketching, measurement, shapes, tools, problem solving.

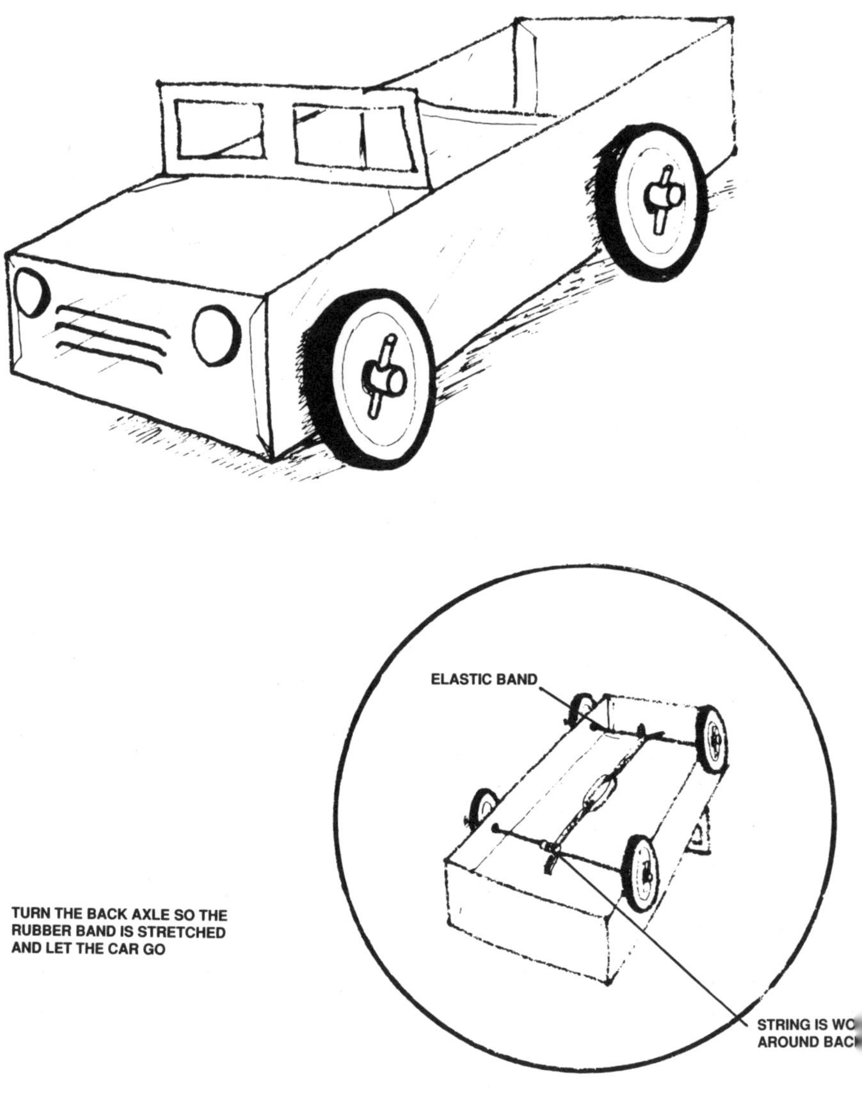

Fig. 12.8 Elastic-powered toys

ELASTIC-POWERED TOYS Fig 12.8

Subjects covered Materials, measurement, TD, sketching, design, tools, simple machines.

Note Many variations of elastic-powered toys are possible. It is important to experiment.

Fig. 12.9 Figure on a stick

FIGURE ON A STICK Fig. 12.9

Subjects covered Materials, measurement, tools, joining techniques, sketching, marking out, drilling, simple mechanics.

T-SQUARES

Subjects covered Wood, measurement, 90° angle, woodwork, tools, glues/nails/screws, TD, marking out, sawing.

SET SQUARES

Subjects covered Wood or metal tools, angles, TD.

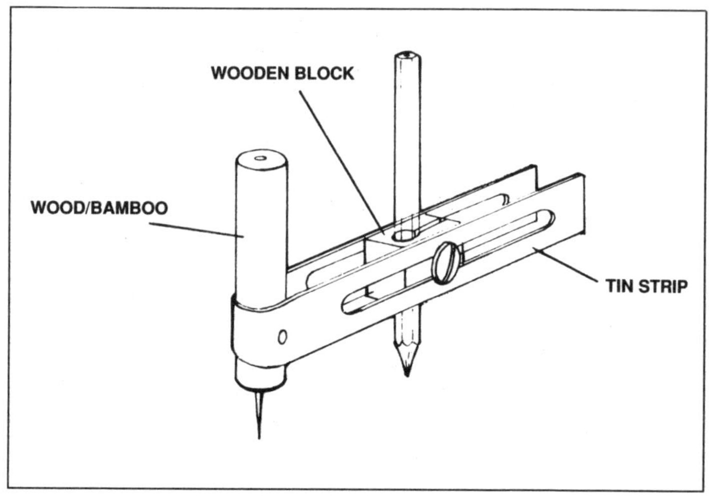

Fig. 12.10 Compass

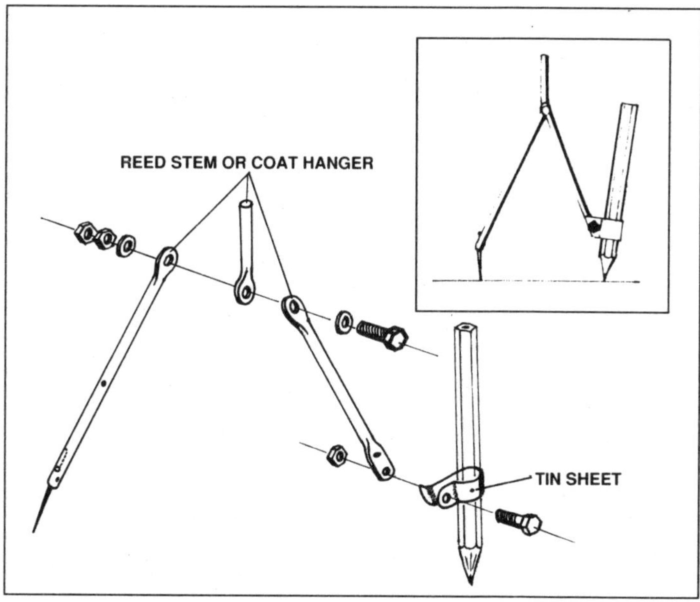

Fig. 12.11 Compass

COMPASS Figs 12.10, 12.12

Subjects covered Measurement, tools, joining techniques, sawing, bending, drilling.

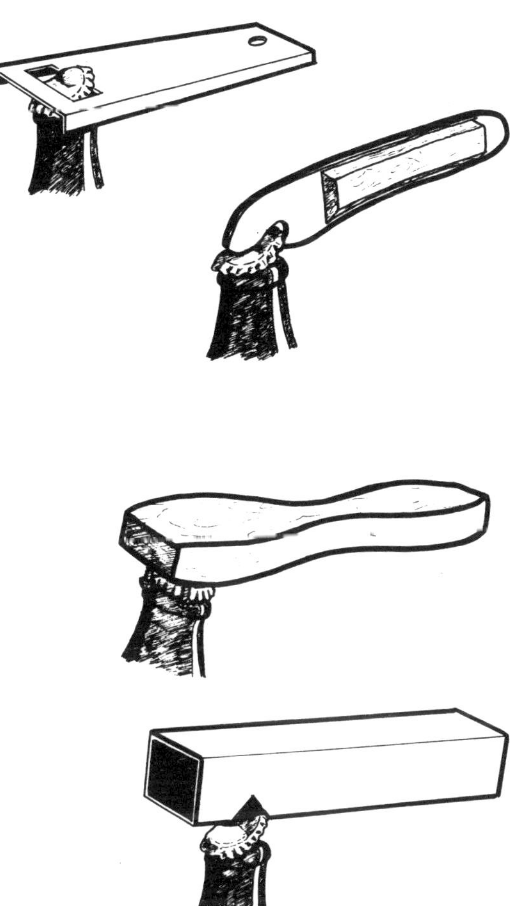

Fig. 12.13 Bottle-openers

BOTTLE OPENERS Fig. 12.13

Subjects covered Measurement, marking out, use of tools, drilling, bending.

Note Many different styles can be made requiring different materials and processes.

MONEY BOXES

Subjects covered Measurement, marking out, tools, joining techniques, screws, nails, rivets, solder, TD.

Note Many different styles can be made from simple, first projects to more complex third-year ones.

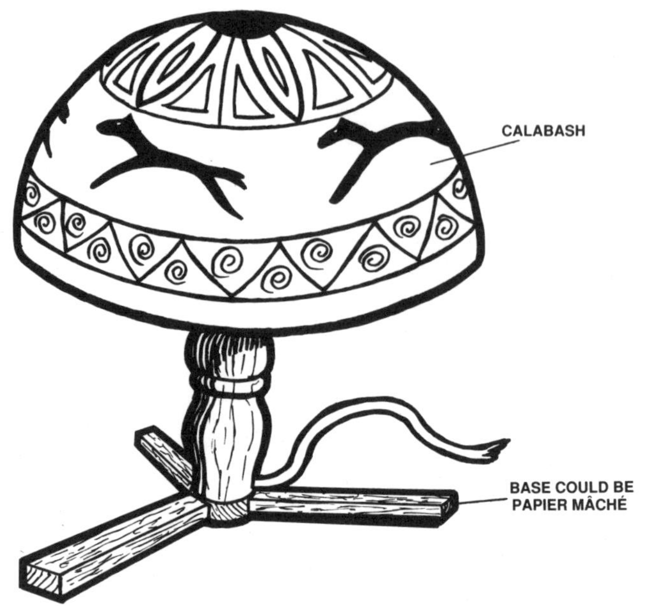

Fig. 12.14 Electric lamps

Fig. 12.15 Electric lamps

ELECTRIC LAMPS Figs 12.14, 12.15

Subjects covered Measurement, tools, design, basic wiring, joining, materials, TD.

Note A calabash makes an excellent lampshade.

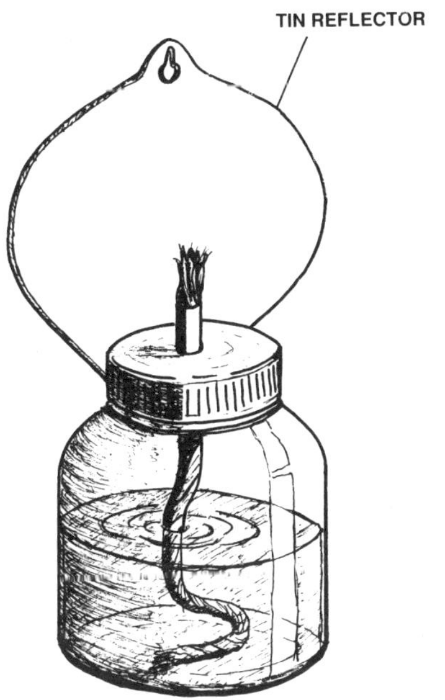

Fig. 12.16 Kerosene lamp

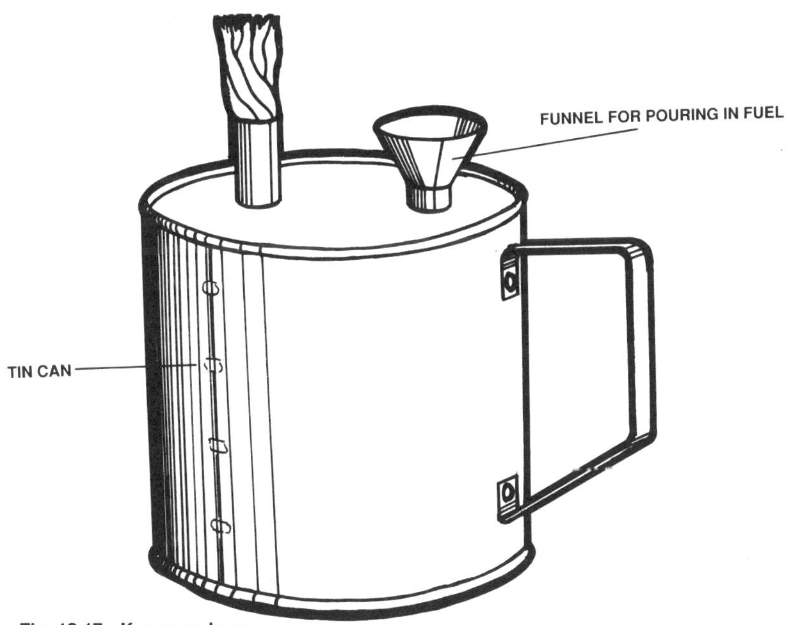

Fig. 12.17 Kerosene lamp

KEROSENE LAMPS

Subjects covered Measurement, metal, tools, soldering, heat, energy, tin-smithing.

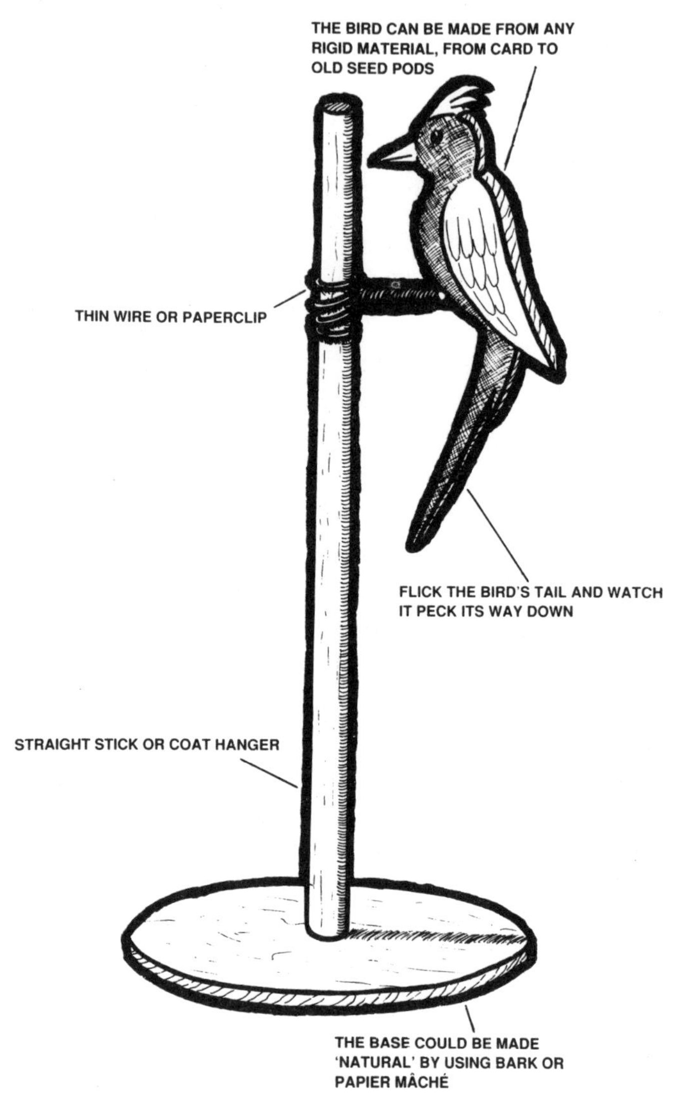

Fig. 12.8 Woodpecker on a tree

WOODPECKER ON A TREE Fig. 12.18

Subjects covered Friction, working properties of materials, design, finishing, measurement, tools, glues, abrasives, joining techniques.

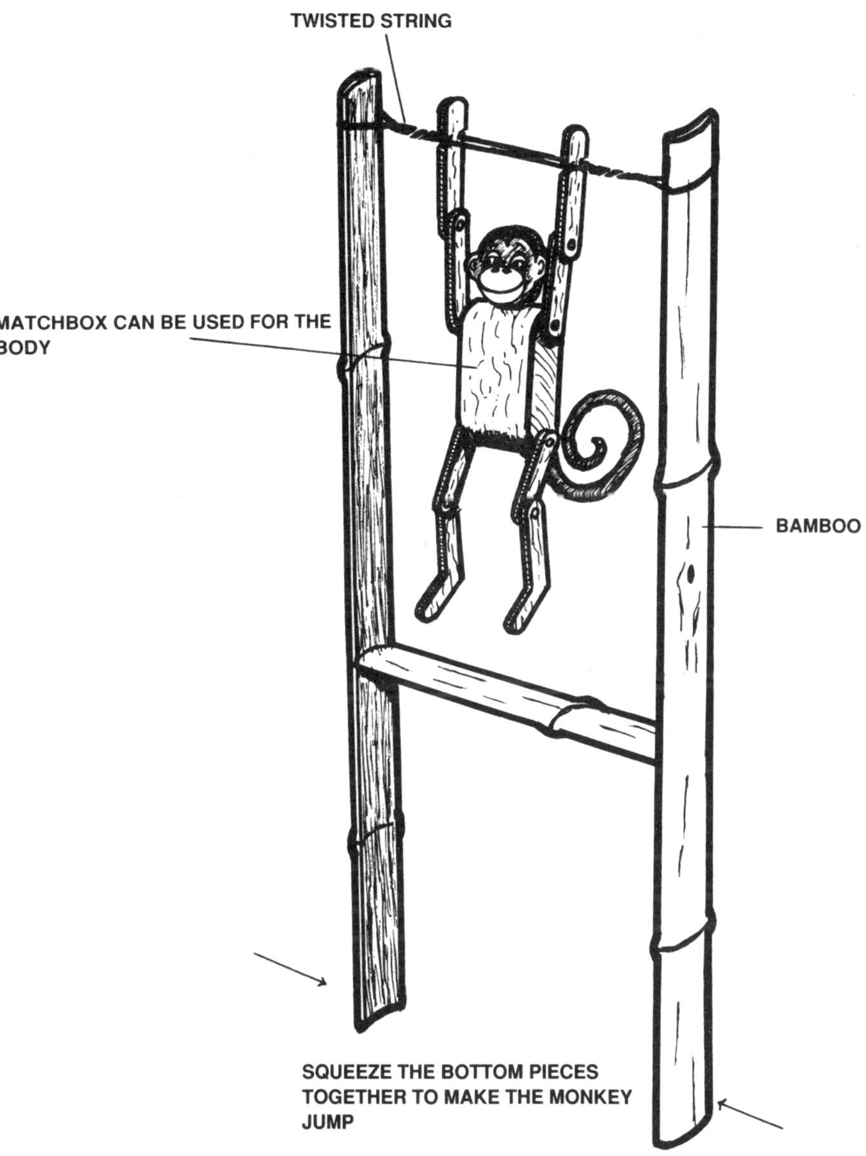

Fig. 12.19 Squeezy acrobatic toy

SQUEEZY ACROBATIC TOY Fig 12.19

Subjects covered Uses of materials, friction, joining techniques, design, measurement, abrasives, energy.

Notes The drawing shows just one method of making the squeezy acrobatic toy. There are many variations in both design and materials.

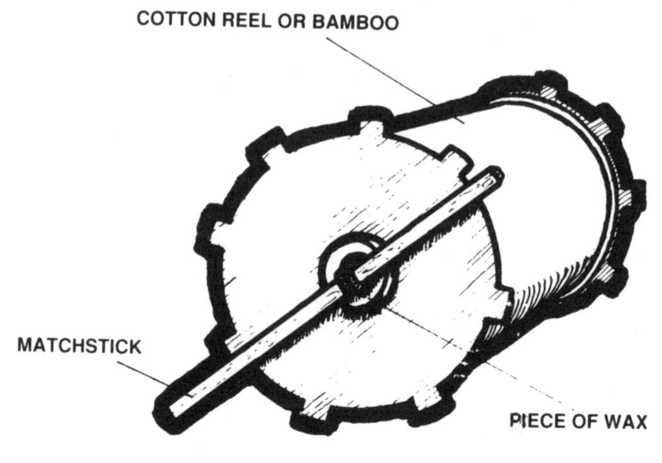

A

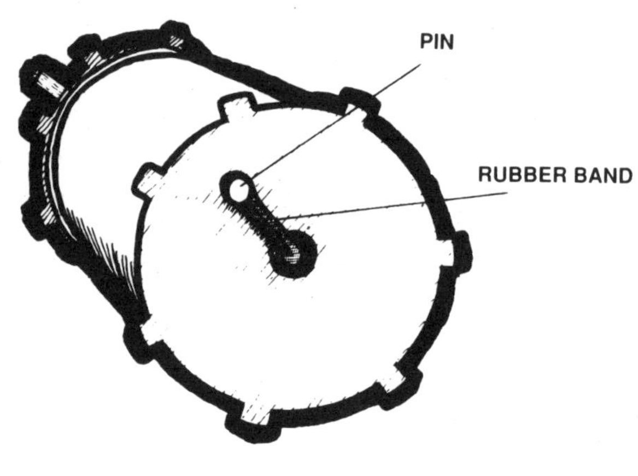

B

Fig. 12.20 Cotton-reel 'tank'

COTTON-REEL CAR Fig 12.20

Subjects covered Simple machines, friction, materials, measurement, tools, power, problem solving.

Notes How it works: place the wax over the hole, fasten the elastic around the stick (A) and pass the elastic through the hole and fasten as in (B). Wind up the stick, put the tank down and watch it move!

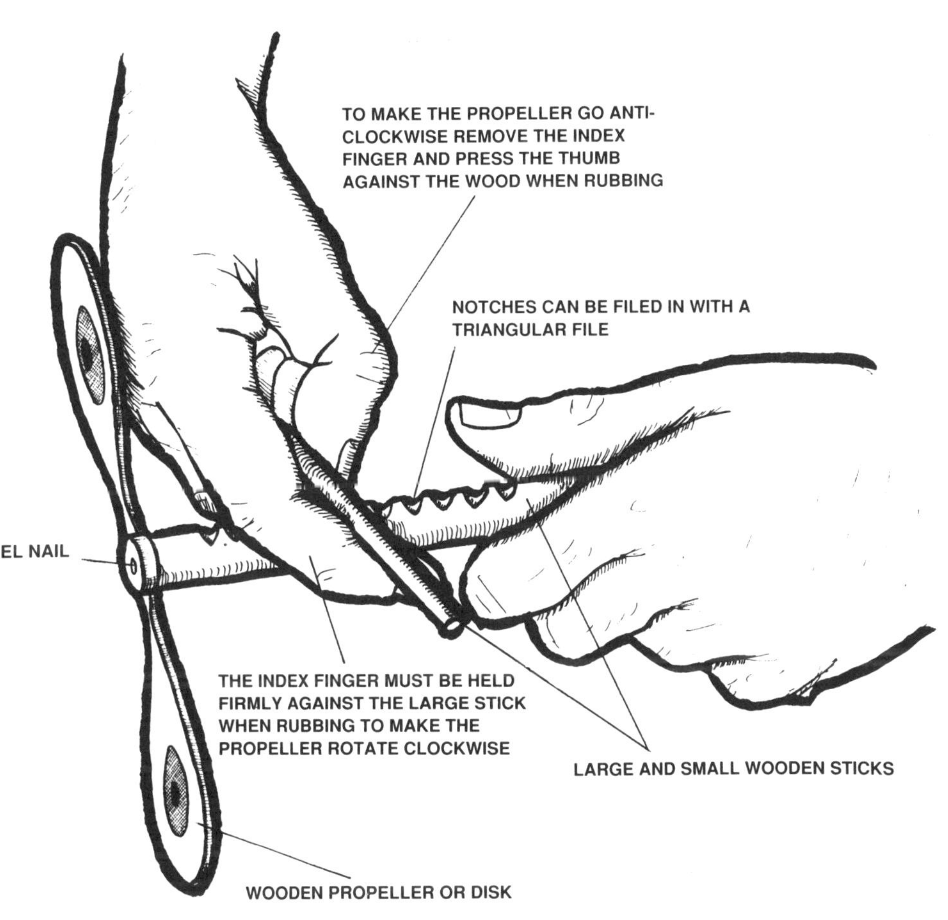

Fig. 12.21 Magic stick

MAGIC STICK Fig 12.21

Subjects covered Materials, properties of metals, vibration, momentum

Notes The small nail used to hold the propeller must not be copper or brass or the magic stick will not work. (Why?)

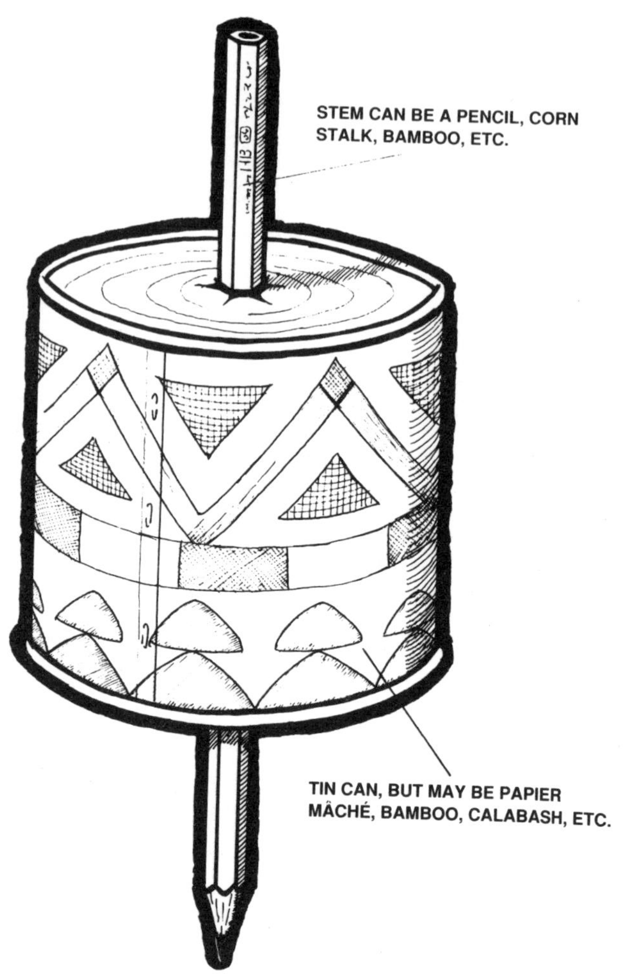

Fig. 12.22 Spinning top

SPINNING TOP Fig 12.22

Subjects covered Measurement, design, tools, energy.

Notes Students can design patterns that will look attractive when the top spins.

YOYO

Subjects covered Measurement, tools, materials, motion, design, finish, joining techniques.

Notes

A small yoyo can be made from two large buttons. If making the yoyo from wood the hole must be central.

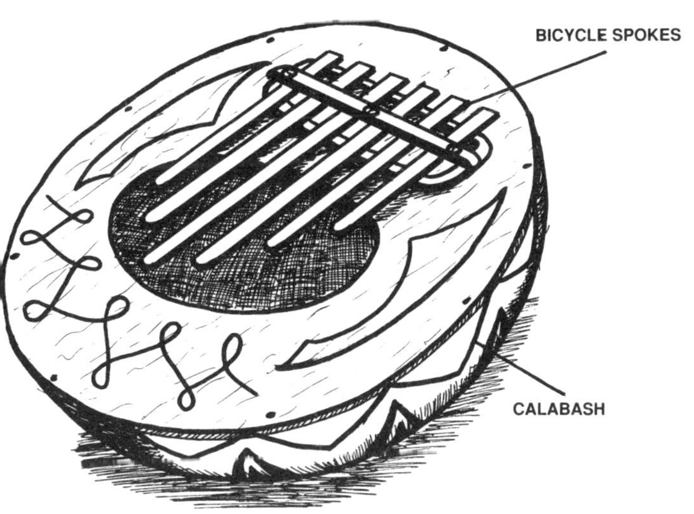

Fig. 12.23 Thumb piano

THUMB PIANO Fig 12.23

Subjects covered Materials, measurement, tools, properties of materials, design, finishes, joining techniques.

Notes The sound board must be as thin as possible.

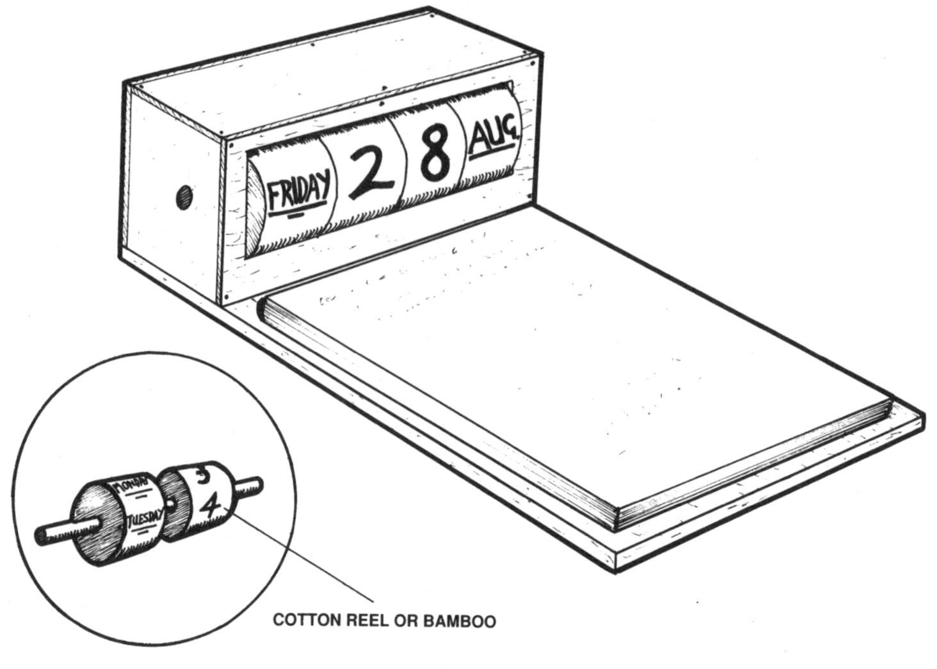

Fig. 12.24 Calendar

Fig. 12.25 Calendar

CALENDAR Figs 12.24, 12.25

Subjects covered Materials, tools, measurement, finish, sheet-metal processes, problem solving (finding out which numbers go where on the blocks).

Notes

The numbers are written as follows: 0, 6, 4, 2, 5, 1 on one block and 2, 0, 7, 1, 8, 3 on the other.

Fig. 12.26 Pencil holder

Fig. 12.27 Pencil holder

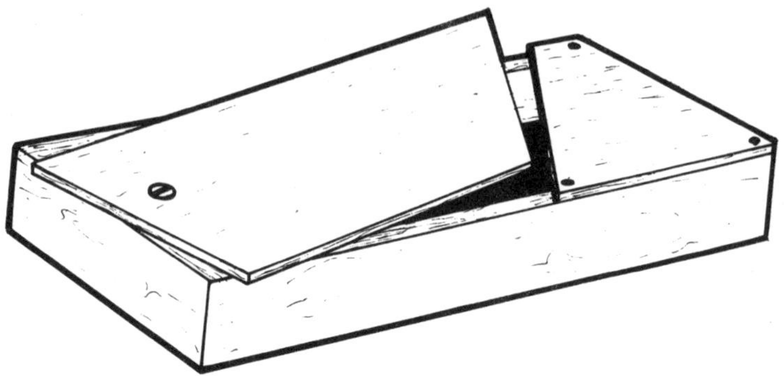

Fig. 12.28 Pencil box

PENCIL HOLDERS AND BOXES Figs 12.26, 12.27, 12.28

Subjects covered Materials, drilling, design, measurement, tools, finishing.

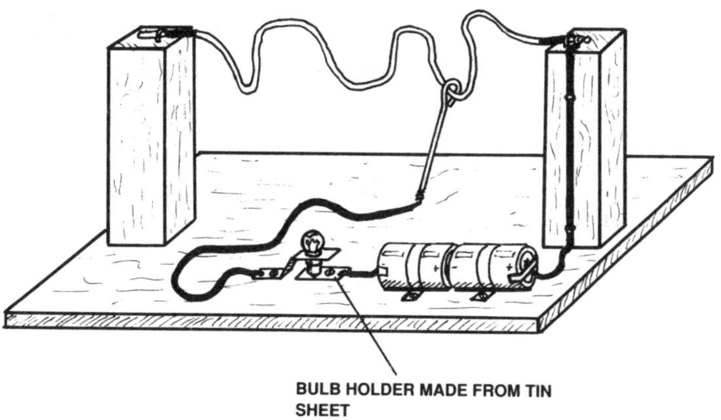

Fig. 12.29 Steady hand game

STEADY HAND GAME Fig. 12.29

Subjects covered Electrical materials, tools, design, measurement, finishing.

Notes Small games can be made by using a medium-sized tin as the base (batteries go into the tin). Simple switches and bulb holders are easily and cheaply made (Chapter 11).

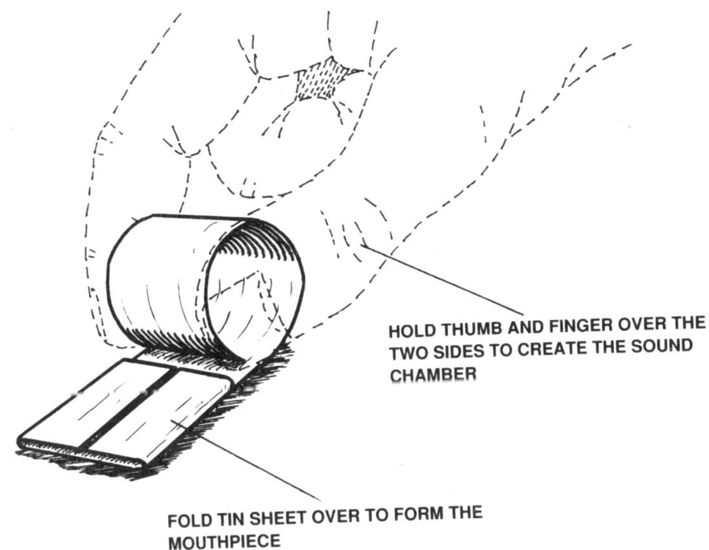

Fig. 12.30 Tin whistle

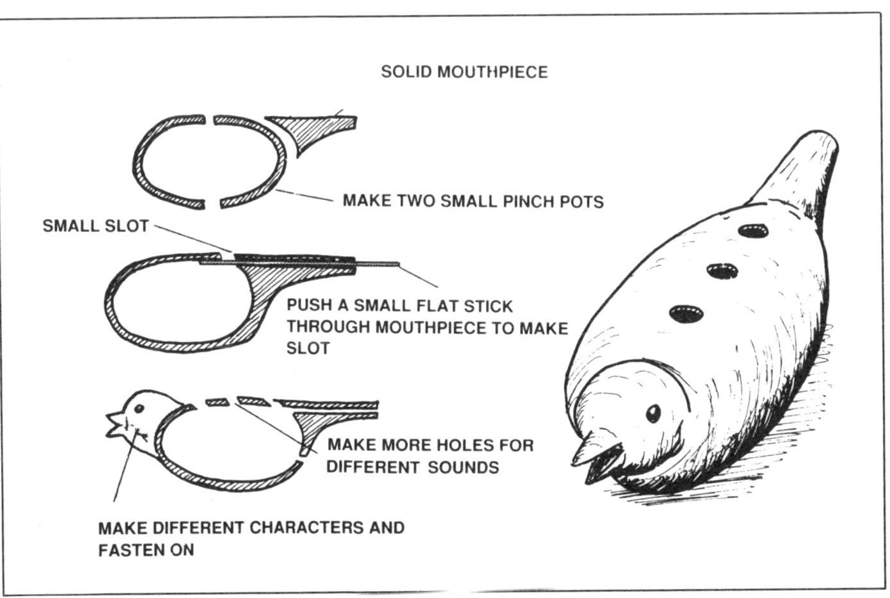

Fig. 12.31 Clay whistle

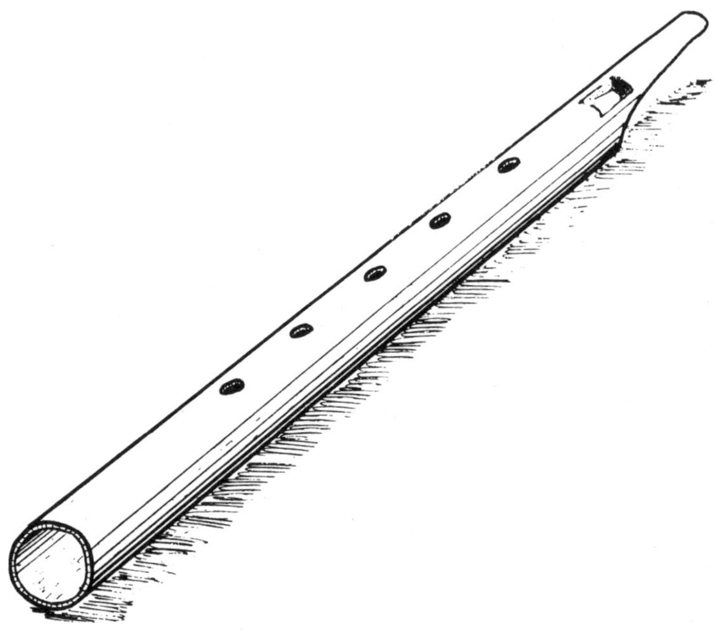

Fig. 12.32 Whistles

TORCH

Subjects covered Electrical, materials, measurement, design, finishing, tools, joining techniques.

Notes Bamboo or tin can be used for the body, clear plastic for the lens. A papier-mâché body could be formed around a frame and inner tube wrapped round afterwards.

WHISTLES Figs 12.30, 12.31, 12.32

Subjects covered Tools, materials, design, joining techniques, finishing, measurement.

Note Most areas have traditional instruments – find out if whistles are among these.

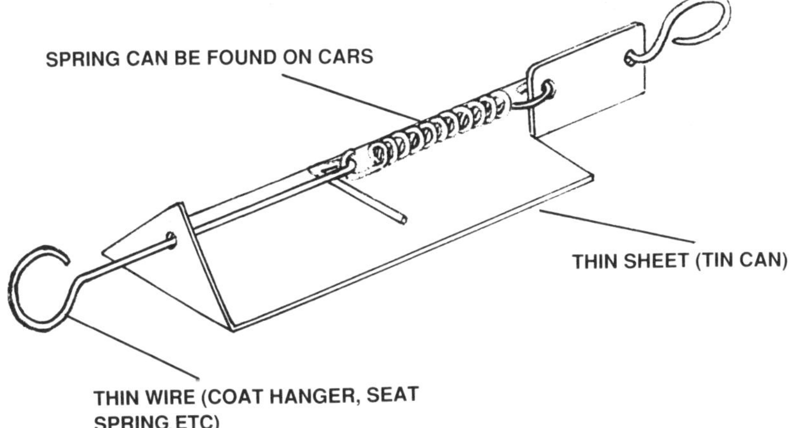

Fig. 12.33 Weighing scales

WEIGHING SCALES Fig. 12.33

Subjects covered Tools, materials, joining techniques, friction, force, energy.

Notes Many different types of scales can be made, from small spring balances to larger cooking scales. Very sensitive scales can be made by using rubber instead of a spring. Set students the problem of discovering how to calibrate the scales.

SANDALS

Subjects covered Tools, materials, joining techniques, measurement, friction, design, finishing.

Notes Rubber from car tyres can be used for the base while raffia or similar is attractive for the thongs.

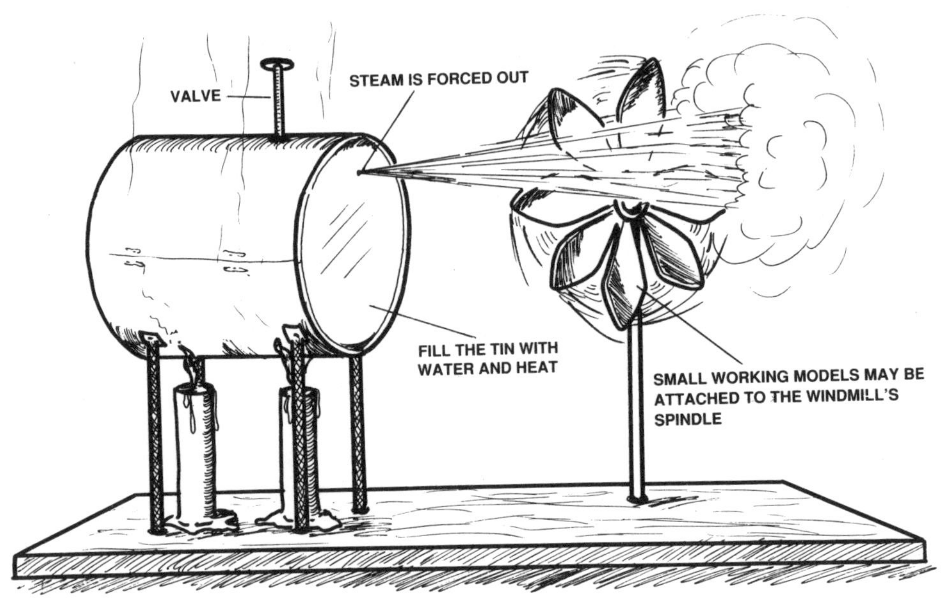

Fig. 12.34 Steam engine

STEAM ENGINE Fig. 12.34

Subjects covered Energy, metalwork, tools, design, finishing.

Notes This is an extremely simple model of a steam engine but shows the principle of steam power very well.

TABLE TENNIS BATS

Subjects covered Woodwork, joining techniques, tools, design, measurement, finishing.

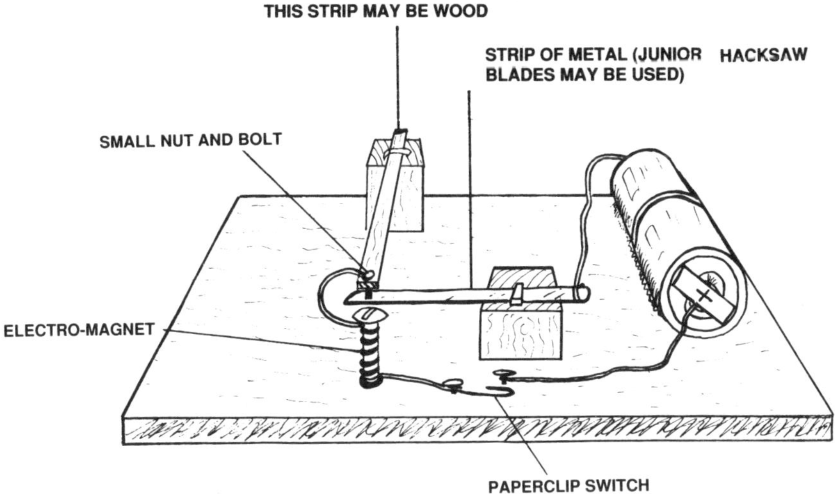

Fig. 12.35 Electric buzzers

ELECTRIC BUZZERS Fig. 12.35

Subjects covered Electricity, materials, tools, design.

Notes When current flows the metal strip vibrates.

DOMINOES

Subjects covered Tools, measurement, materials, maths, finishing.

Notes You need 28 dominoes for the standard set. However, it is well worth designing your own games using dominoes maybe with shapes or numbers.

TOOLS

Subjects covered Measurement, design, tools, application of materials, joining techniques, basic machines.

Notes The making of tools is a great way of linking together all technical subjects. For example, if students are making a wooden calendar they could make very simple metal tools for carving the numbers and names of the months. Very simple tools can be made for decorating clay work.

Fig. 12.36 Puppets

PUPPETS Fig. 12.36

Subjects covered Materials, motion, design, measurement, tools.

Notes Very simple, clay finger-puppets make a simple first project.

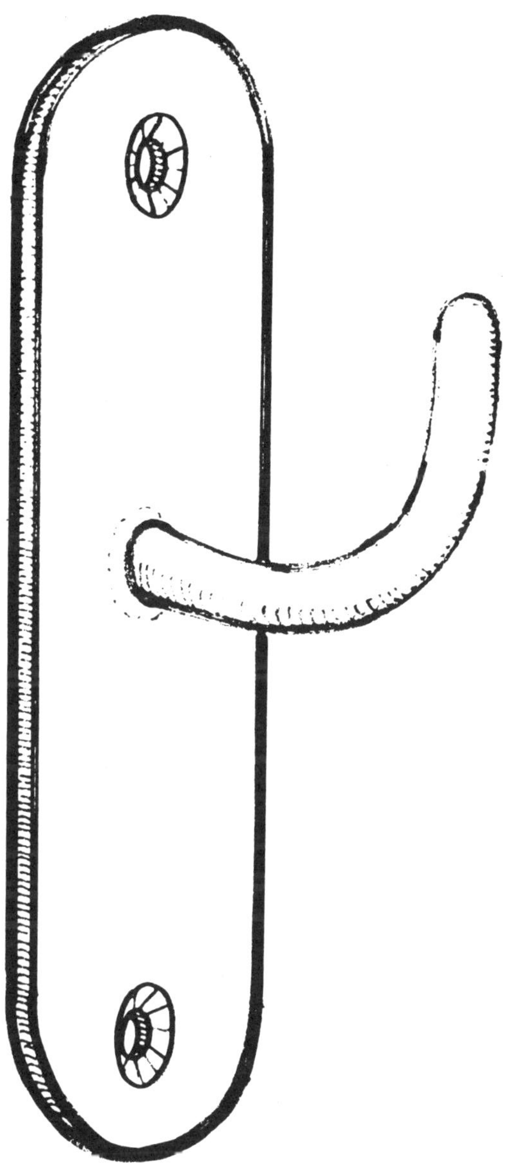

Fig. 12.37 Hooks and hangers

HOOKS AND HANGERS Fig. 12.37

Subjects covered Materials, design, measurement, tools, finishing, joining techniques.

ACTIVITIES

The following is a list of activities that have proved very efficient in the teaching of introductory technology. The activities are listed under subject headings for convenience.

MEASUREMENT

For teaching basic measurement, get the students to pace out the room, football pitch, etc. From the results the students will start to learn about standards. Once students are all using the same standard they can begin to measure personal details such as height, handspan, etc. Don't forget other forms of measurement such as time and temperature. From here on, measurement should constantly be reinforced.

MATERIALS

'Materials' may be started by asking students to bring in as many different materials as they can find. From the assortment of materials, students can start to discover about different properties through experiment (ask why we cannot use clay for a hammer or make a pan from wood; very simple but it gets the use of materials across well). Identify materials through their properties.

Brittleness Hit different materials with a hammer to discover which materials are most brittle. Explain malleability at the same time.

Conductivity Use a circuit to discover which materials conduct electricity (be sure to use pencil lead as one example – it always amazes students). Put materials into hot water to see which ones conduct heat (ask students what they can use the poor conductors for).

Weight Measure weight by comparison; use a simple seesaw to compare weights.

Sound Percussion test; students to close their eyes and identify the material from the sound.

Ferrous/non-ferrous Use magnets; make a fishing game where a magnet is attached to a piece of string and students have to pick up cardboard fish which have different materials attached to them, i.e., paperclip (ferrous), aluminium (non-ferrous).
Show different items using two or more materials and ask students why the particular materials have been chosen, e.g. plastic bucket with a metal handle, catapult of wood and rubber.

MAINTENANCE

Use pepper grinders, stoves, lamps and bicycles to demonstrate maintenance. Show students the importance of being methodical when removing parts, and of keeping them clean.

ENERGY

Show solar energy by setting fire to a piece of paper using a magnifying glass (stress safety). To show wind energy make kites, paper windmills, sailboats; demonstrate chemical energy by striking a match, using kerosene lamps or a stove and showing a tyre-vulcanizer. Have a 100-yard sprint to demonstrate energy; ask students where their energy source comes from (food). Students to produce a list detailing sources of energy, the type of energy produced and whether it is exhaustible or not.

FRICTION AND FORCE

Show force by having arm-wrestling competitions and tug-of-war games. Build a slope from a piece of metal or glass. Put different materials at the top of the slope and see which ones slide best. (Why?) Put oil on to the slope and see what happens. Find a container with a screw top and ask students to unscrew it. Put Vaseline on the top and repeat the experiment. Rub hands together and feel the heat created. Find examples of friction in everyday use, e.g. brakes, striking a match.

LEVERS

Set the problem of moving a heavy object with only a large stick to help. Build a seesaw, look at everyday items and discover which ones use levers.

PULLEYS AND GEARS

Students to make pulleys and stands from available materials and experiment with lifting different items. Make simple gears from corrugated cardboard and coke-bottle tops.

ELECTRICITY

Static electricity: suspend a plastic pen from a piece of string (so the pen hangs horizontally). Get a variety of objects made from plastic, glass, wood and metal and some different kinds of materials such as cotton, wool, nylon and fur. Rub each object with one of the materials and bring it close to the pen. An object with a charge makes the pen swing round. If balloons are available they can be 'electrified' and used to pick up pieces of paper.

Batteries: take old dry-cell batteries apart to find out what's inside. Make wet-cell batteries from lemons: stick a piece of copper and a piece of zinc into a lemon, attach wires to each piece of metal, then touch them on your tongue.

To show that electricity produces heat, carry out the following experiment: wind a thin strand of wire wool around a nail to form a spring, remove the nail and attach the wires from a battery to the two ends of the wire wool; the wool will glow red. By repeating the experiment with a very short piece of wire wool, students can see how a bulb works.

Electro-magnets: wind very fine copper wire (old motors provide a good source) around a nail and attach the two ends of the wire to the battery.

BUILDING, SANITATION AND WATER

Make water filters by cutting the bottom off a plastic bottle; turn the bottle upside down and fill with coarse gravel, fine gravel and sand, then pour in dirty water; the water will be clear when filtered through.

Brickmaking: the majority of building should be taught through a building project (ask the Agriculture Department if they need any animal houses building); it may mean asking local builders to allow the students to 'help' on a nearby building.

FOOD PRESERVATION

Make a study of local food storage – why are different foods stored in particular ways? Study water-pot fridges (show application of materials very well) – why do they work?

BIBLIOGRAPHY

Ashworth, A.E. (1983, 1986, 1988), *Introductory Technology Drawing*, Books 1–3 (Evans).

Ashworth, A.E. (1982), *Testing for Continuous Assessment – A Handbook for Teachers in Schools and Colleges* (Evans).

Caborn, C. and Mould, I. (1985), *Integrated Craft and Design* (Nelson).

Cohen, L. and Mamion, L. (1989), *A Guide to Teaching Practice* (Routledge).

Geary, K., *Make and Find Out* Series (Macdonald).

Godwin, T. and Wright, M. (1984), *Educational Wooden Toys in Sri Lanka* (ITP)

Harvey, D. (1983), *Imaginative Pottery* (Black).

Introductory Technology, Books 1–3 (Longman).

Ladybird Books on Metalwork, Electricity, Mechanics, and Woodwork.

Lye, P.F. (1985), *Metalwork Theory*, Books 1–4 (Nelson).

Lye, P.F. (1985), *Woodwork Theory*, Books 1–4 (Nelson).

Mamion, L. and Cohen, L. see under Cohen, L. and Mamion, L.

Marland, M. (1975), *Craft of the Classroom – A Survival Guide* (Heinemann Educational).

Moore, A. (1987), *How to Make Planes, Cramps and Vices* (IT Publications)

Moore, A. (1986), *How to Make Twelve Wood-Working Tools* (IT Publications)

Mould, I. see under Caborn, C. and Mould, I.

Oyelola, P. (1983), *Nigerian Crafts* (MacMillan Educational).

Pettit, L., *Appreciation of Materials and Design* (Arnold).

Platt, A. (1986), *Using Wood and Metal in the Workshop* (Evans).

Primmer, L., *Pottery Made Simple* (W.H. Allen).

UNESCO (1982), *Systems Approach to Teaching and Learning Procedures: A Guide for Educators in the Third World* (HMSO).

Wallace, R. (1985), *Introductory Technology*, Books 1–3 (MacMillan Educational).

Ward, A. (1989), *1000 Ideas for Primary Science* (Hodder & Stoughton).

Weygers, A. (1973), *Making of Tools* (Van Nostrand Reinhold).

Weygers, A. (1974), *The Modern Blacksmith* (Van Nostrand Reinhold).

Wright, M. see under Godwin, T. and Wright, M.

The following books are available in Nigeria:

Adeyemo, P.O., *Practice of Education* (Omolayo Standard Press).

Adeyemo, P.O., *Principles of Education* (Omolayo Standard Press).

Introductory Technology Exam Made Easy, Books 1–3 (Book Representation and Publishing Company, Ibadan).

Introductory Technology, Books 1–3 (University Press, Ibadan).

Introductory Technology for Schools and Colleges, Teacher's Books 1–3 (Evans, Nigeria).

RESOURCES

Apart from the books mentioned in the bibliography there are many other useful sources of information which will be of great help to teacher and students alike.

NEWSPAPERS

In Nigeria a monthly journal for schools and colleges called *Young Talent* has been introduced, which is an excellent paper. The address is: *Young Talent*, P.O. Box 1661, Somolu, Lagos State, Nigeria. In other countries, similar journals may be available. Also, other papers are very good for collecting pictures of technology.

CALENDARS

Look around at all the calendars – some have very good pictures of technical items.

POSTERS

Some states have produced good-quality posters of tools. Ask at your Ministry of Education if they are available.

LARGE COMPANIES

It's well worth writing to large companies such as the oil, mining and sawmill companies to see if they have any literature about their industries. If you visit them they may be able to give you material samples.

JELLY DUPLICATOR

Duplicating kits containing special jelly mixes and pens can be bought. These kits are very efficient and have been used with great success all over the world. An address for supply is:

Hecto Duplicator Co.
Mawdesley,
Ormskirk,
Lancs,
L40 2RL, UK

CENTRE FOR ALTERNATIVE TECHNOLOGY

CAT has an excellent display of alternative technology; it also has a very comprehensive bookshop (highly recommended):

CAT
Llwyngwern Quarry,
Machynlleth,
Powys,
Wales, UK.

TALC (TEACHING AIDS AT LOW COST)

Suppliers of a large range of books and slides for teachers and health workers with a small budget. Titles include *Toys for Fun* and *Simple English is Better English.*

P.O. Box 49,
St. Albans,
Herts AL1 4AX, UK

THE IT BOOKSHOP

Stocks a wide range of practical and technical books for development workers. A mail order catalogue *Books by Post*, listing over 650 key titles, is also available:
103–105 Southampton Row,
London WC1B 4HH, UK

EDUCATION RESOURCE CENTRES

Many towns have resource centres that will have reference materials and often professional support. (Ask the local ministry as these centres are often a well kept secret!)

THE ECOE PROGRAMME

(Evaluating and Communicating our Overseas Experience)

THE NEED

Over the past thirty years, more than 25,000 volunteers have worked abroad with VSO. Currently, there are over 1,200 volunteers working in over 40 developing countries in Africa, Asia, the Pacific and the Caribbean for periods of two years or more. However, we have become increasingly aware that much of this valuable experience has been lost through not being recorded in ways which make it accessible and communicable. The ECOE Programme addresses this problem.

THE AIM

The aim is to record volunteers' experience in reports, videos, seminars, conferences, books, etc. This body of knowledge supplements and supports the work of individual volunteers. It also provides information which is accessible not only to volunteers but also to their employers overseas and to other agencies for whom the information is relevant. Care is taken to present each area of volunteer experience in the context of current thinking about development so that VSO both contributes to development discussions and learns lessons from them for the continuance of its work.

ADVISORY PANEL

A panel of opinion leaders in relevant professions and in development thinking advises on the selection and commissioning of ECOE publications.

For further information write to:

The Programme Evaluation Manager
VSO
317 Putney Bridge Road
London SW15 2PN, UK

Tel: 081-780 2266 Fax: 081-780 1326